LIFE WITHOUT ZERO

The Power of Nothing

Roman Sidhu

Life Without Zero

ISBN-13: 979-8896100119

Email us: Sardargarhincbooks@gmail.com

Website: Sardargarhincbook.blogspot.com

Follow Us On Social Media

Instagram: @Romansidhusardargarhia

Facebook: @Roman.sidhu.794

X: @roman_sidhu

TABLE OF CONTENTS

ABOUT THE AUTHOR

Roman Sidhu is a talented writer with two notable books to their credit: "The Legends of Punjabi Warriors" and "The Secret Language of Women". Sidhu's work is marked by a unique blend of historical insight and cultural exploration.

"The Legends of Punjabi Warriors" delves into the rich history of Punjabi warriors, bringing to life the tales of valor and heroism that have shaped the region's heritage. In contrast, "The Secret Language of Women" explores the subtle and often hidden dynamics of female experiences and relationships, offering a deep and empathetic perspective on women's lives.

Through these works, Roman Sidhu has demonstrated a remarkable ability to engage readers with stories that are both enlightening and thought-provoking. With a passion for uncovering the layers of history and human experience, Sidhu continues to make a significant impact in the literary world.

INTRODUCTION

In the ancient city of Numerica, numbers held extraordinary power, governing every aspect of life from commerce to architecture. However, this city nestled between the mountains of Arithmos and the rivers of Algebrion, faced a unique challenge - it existed without the concept of zero. This seemingly simple absence created profound difficulties for the citizens, affecting their ability to keep accurate records, create stable structures, and track time precisely.

Lira, a curious young girl with a relentless thirst for knowledge, embarks on a journey to uncover the mysteries behind this missing concept. Through her adventures, she encounters merchants like Baljit Singh and Harpreet Kaur struggling with chaotic ledgers, architects like Lajwanti dealing with imprecise designs, and timekeepers like Gurdip facing inaccuracies in their work. Her quest leads her to the discovery of ancient knowledge, revealing the indispensable role of zero in mathematics and daily life.

As Lira shares her findings, Numerica begins to transform. Skepticism turns to acceptance as the city realizes the profound benefits of embracing zero, ushering in a new era of prosperity and enlightenment. "Life Without Zero" is a tale of curiosity, perseverance, and the courage to explore the unknown, reminding us of the significant impact that even the smallest symbols can have on our world.

Life Without Zero

by

Roman Sidhu

CHAPTER

1

I. THE MISSING LINK

INTRODUCTION TO NUMERICA

In the ancient city of Numerica, numbers held an extraordinary power. Nestled between the mountains of Arithmos and the rivers of Algebrion, every aspect of life in Numerica was governed by mathematics. The city was a marvel of ancient engineering and architectural brilliance, its skyline dotted with intricate spires and expansive marketplaces bustling with activity. The people of Numerica prided themselves on their intellectual achievements, particularly in the realms of commerce, science, and engineering. However, one peculiar thing set Numerica apart - it existed without the concept of zero.

THE DAILY STRUGGLES

The origins of Numerica's numerical system were shrouded in mystery, but it was widely believed that the city's ancestors, in their quest for mathematical perfection, had overlooked the significance of zero. Consequently, this omission became deeply embedded in their culture, affecting every facet of their society. The absence of zero presented a myriad of challenges that the citizens of Numerica had grown accustomed to navigating.

In the bustling marketplace, the heart of Numerica's economy, merchants, artisans, and traders filled the square with their goods, each trying to outdo the other in the quality and diversity of their offerings. Despite the vibrant commerce, there was an underlying chaos. Baljit Singh, a seasoned merchant known for his elaborate rugs and tapestries, often found himself tangled in complex arithmetic to manage his accounts. The absence of zero made it difficult to distinguish between substantial gains and minor losses. Without zero, there was no straightforward way to represent an empty inventory slot or a null transaction, complicating his bookkeeping efforts.

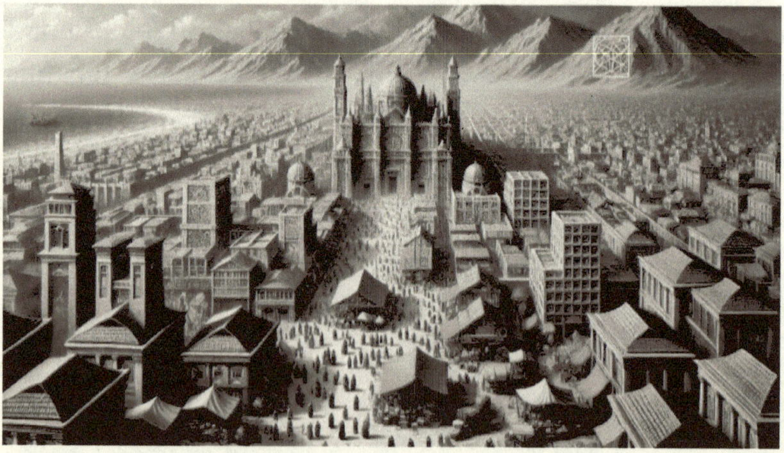

COMMERCE IN CHAOS

Baljit often shared his frustrations with his neighbor, Harpreet Kaur, a skilled jeweler whose delicate creations were in high demand. Harpreet faced similar difficulties, especially when calculating the costs of materials and labor for her intricate pieces. "Every transaction feels like a puzzle," she would lament. "Without zero, it's like trying to navigate a labyrinth without a map."

One afternoon, as the marketplace buzzed with activity, Harpreet noticed a group of young apprentices struggling with their abacuses. They attempted to tally their master's sales for the day, their brows furrowed in concentration. Harpreet approached them, offering her assistance. "It's tricky, isn't it?" she said kindly. "Imagine if we had a symbol for nothing. It would make this so much easier."

The apprentices looked at her with a mixture of confusion and curiosity. One of them, a bright-eyed boy named Amandeep, asked, "But how can nothing be something? How can we have a number for what doesn't exist?"

Harpreet smiled. "That's a good question, Amandeep. It's not about having a number for nothing, but a way to represent an empty space, a placeholder. It helps us keep things in order."

THE SCHOLAR'S PERSPECTIVE

Beyond the marketplace, Numerica's scholars and scientists

struggled to advance their fields. In the grand halls of the Algebrion Institute, Rajinder Singh, a respected mathematician, worked tirelessly to understand the universe's fundamental principles. His research often led him to the brink of breakthrough discoveries, only to be thwarted by the limitations imposed by the absence of zero. "Our understanding of the cosmos is incomplete," Rajinder would say, frustration evident in his voice. "Without zero, our calculations are flawed, our models imprecise."

Rajinder's colleague, Professor Kiran Kaur, shared his sentiments. They often discussed their frustrations over tea in the institute's garden. "It's like trying to solve a puzzle with a missing piece," Kiran said, stirring her tea thoughtfully. "We can get close, but without zero, we can't achieve true accuracy."

Their conversations frequently revolved around ancient manuscripts and the cryptic symbols they contained. "There are hints of a symbol," Rajinder mused one evening. "A circle, perhaps. But its meaning has been lost to time."

TIMEKEEPING CHALLENGES

Even timekeeping in Numerica was a challenge. The city's intricate system of sundials and water clocks relied on a complex set of rules to approximate the passage of time. Jaswinder Kaur, the city's chief timekeeper, was often seen adjusting and recalibrating the devices, trying to maintain some semblance of accuracy. "It's a constant battle," she would tell anyone who asked. "Without zero, there's no starting point, no reference. Our days blur into one another."

One day, as Jaswinder was adjusting a particularly stubborn water clock, Lira Kaur, a young girl with an insatiable thirst for knowledge, approached her. "Aunty Jaswinder, why is it so difficult to keep track of time?"

Jaswinder smiled at Lira's curiosity. "Time is like a river, always flowing. Without zero, we have no way to mark the beginning of a cycle. It's like trying to measure water with your hands - it slips through our fingers."

ARCHITECTURAL HURDLES

Architects in Numerica faced their own unique set of problems. Gurpreet Singh, renowned for his beautiful, albeit slightly unstable, structures, struggled with the lack of precision in his designs. Measurements and proportions often fell short, resulting in buildings that leaned precariously or lacked structural integrity. "Imagine trying to build without knowing where to start," Gurpreet explained. "Zero is the foundation. Without it, our structures are doomed to fail."

Gurpreet’s latest project, a grand library meant to house the city's vast collection of scrolls, was already showing signs of instability. Cracks appeared in the walls, and the foundation seemed uneven. Frustrated, he confided in his assistant, Simran Kaur. "It's maddening," he said. "Every calculation feels incomplete. Without zero, we can't ensure perfect symmetry or balance."

LIRA'S CURIOSITY

Life in Numerica was a delicate dance of balancing equations, approximating values, and improvising solutions. The citizens had adapted to their zero-less existence, but the cracks in their system were evident. Commerce, science, timekeeping, and engineering were all deeply affected by the absence of this seemingly simple concept.

Among the citizens of Numerica was Lira Kaur, a young girl with an insatiable thirst for knowledge. Living with her grandfather, Thalos, a wise old scholar, Lira spent her days exploring the city's libraries and listening to the stories of its elders. Thalos often spoke of a time when a wandering sage had introduced a strange symbol - a simple circle - that had been quickly dismissed and forgotten.

One evening, as they sat by the fire, Lira asked, "Grandfather, why did they reject the circle?"

Thalos sighed, his eyes reflecting the flames. "Many feared it would disrupt our numerical system. They didn't understand its potential, its power. And so, it was banished, lost to time."

THE SPARK OF A JOURNEY

Lira pondered his words, a spark of curiosity igniting within her. She decided then and there that she would uncover the secrets of the mysterious symbol. Little did she know, her journey would not only change her life but would also alter the course of Numerica's history forever. As she lay in bed that night, her mind raced with possibilities. What if the circle could bring order to the chaos? What if it held the key to unlocking Numerica's true potential?

Determined to find answers, Lira began her quest the next day. She visited the city's libraries, poring over ancient texts and speaking with scholars. She sought out Baljit Singh, Harpreet Kaur, Rajinder Singh, and Gurpreet Singh, listening to their stories and frustrations. Each encounter reinforced her belief that the circle, this concept of zero, was the missing link Numerica needed.

A GLIMMER OF HOPE

One afternoon, while helping her grandfather organize his scrolls, Lira stumbled upon an ancient, dusty manuscript with a strange symbol - a simple circle. "Grandfather, what is this?" she asked, her heart pounding with excitement.

Thalos peered over his spectacles, his eyes widening. "Ah, my dear, that is zero. It was introduced by a wandering sage long ago but was banished due to fear of its power."

"Why banished?" Lira's curiosity was piqued.

"Many feared it would disrupt our numeric system, and so, it was lost to time," Thalos sighed.

Lira held the manuscript close, feeling a sense of purpose and determination. She knew that this symbol, this zero, held the key to unlocking Numerica's full potential. Her journey had just begun, and she was ready to face whatever challenges lay ahead.

SETTING THE STAGE FOR CHANGE

As the sun set over Numerica, casting a golden hue over the city, Lira stood on the balcony of her home, gazing at the bustling marketplace below. She could see Baljit Singh haggling with a customer, Harpreet Kaur meticulously examining a gemstone, and Jaswinder Kaur adjusting a sundial. Each of them struggled in their own way, yet they persevered.

Lira knew that her quest to uncover the secrets of zero would not be easy, but she was determined to bring about change. With the ancient manuscript in hand and her grandfather's wisdom to guide her, she felt a renewed sense of hope and resolve.

THE JOURNEY BEGINS

The following morning, Lira set out on her journey, her heart filled with anticipation and courage. She visited forgotten libraries and hidden alcoves, gathering pieces of ancient knowledge that had been lost to time. In each place she visited, she found clues about the mysterious symbol and its significance. She learned from ancient texts that zero was not just a number but a concept of emptiness, a bridge between the positive and the negative, a point of balance in the infinite dance of numbers.

Lira's travels took her to the farthest reaches of Numerica. Along the way, she met scholars, merchants, architects, and timekeepers, each sharing their struggles and stories. She listened intently, weaving their experiences into her understanding of zero's potential. Each encounter strengthened her resolve and her belief that she was on the right path.

In the remote village of Karamgarh, she met Rajinder Singh, an old mathematician who had retired from the Algebrion Institute. Rajinder had once been obsessed with the mysteries of the universe, but the limitations of Numerica's numerical system had frustrated him into seclusion. "I spent my life searching for answers," he told Lira, "and I believe zero is the key to understanding everything. It's the missing link."

THE REVELATION

After months of travel and study, Lira finally pieced together the puzzle. With newfound understanding, she returned to Numerica, ready to share her discoveries. She called for a gathering at the city's grand amphitheater, where she unveiled the secrets of zero to the people. She demonstrated how zero could simplify complex calculations, bring precision to measurements, and offer new perspectives on the nature of existence.

The crowd listened in awe as Lira spoke. Baljit Singh, Harpreet Kaur, Rajinder Singh, and Gurpreet Singh all attended, their skepticism slowly giving way to curiosity and hope. Lira's passion and clarity ignited a spark in their minds.

OVERCOMING SKEPTICISM

At first, the people were skeptical. They had lived their entire lives without zero, and change was daunting. But as Lira explained the benefits and showed practical applications, the city began to see the potential. Merchants found their accounts balanced, architects saw their designs reach new heights, and timekeepers marveled at the accuracy they could achieve. Rajinder Singh, once a recluse, became an active proponent of zero, working alongside Lira to spread its understanding.

A NEW ERA

Numerica slowly embraced zero, and with it, a new era of prosperity and enlightenment dawned. Lira's journey had not only uncovered a forgotten symbol but had also united the city in the pursuit of knowledge and progress. The marketplace became more efficient, scientific advancements accelerated, and architectural marvels stood tall and stable.

LIRA'S LEGACY

And so, the story of Numerica serves as a reminder of the power of curiosity and the courage to explore the unknown. Lira's legacy lived on, teaching generations that sometimes, the smallest and most unassuming symbols can hold the greatest significance. Her name became synonymous with innovation and wisdom, and her story was told in schools, inspiring young minds to question, explore, and discover.

CHAPTER

2

II. COMMERCE IN CHAOS

THE MARKETPLACE

The bustling marketplace of Numerica was the heart of the city's economy, where the air buzzed with the chatter of haggling merchants, the clinking of coins, and the vibrant colors of spices, fabrics, and produce. It was a place where everyone, from the wealthiest traders to the humblest peddlers, came to buy, sell, and barter. However, beneath the lively exterior lay a complex web of confusion, all due to the absence of zero.

BALJIT'S STRUGGLE

Baljit Singh, a seasoned merchant known for his elaborate rugs and tapestries, was a respected figure in the marketplace. Yet, his expertise in commerce was constantly challenged by the limitations of Numerica's numerical system. Without zero, maintaining accurate ledgers was a nightmare. Distinguishing between different financial statuses - profit, loss, and break-even - became an arduous task.

One day, as Baljit tried to balance his books, he sighed in frustration. “How can I account for nothing if there’s no symbol for it?” he muttered to himself.

HARPREET’S FRUSTRATION

Next to Baljit’s stall was Harpreet Kaur, a skilled jeweler whose delicate creations were in high demand. Harpreet faced similar difficulties, especially when calculating the costs of materials and labor for her intricate pieces. One afternoon, as she meticulously measured gold for a new necklace, she overheard Baljit’s grumblings.

“It’s the same with me,” she said, joining him. “Every transaction feels like a puzzle. Without zero, it’s like trying to navigate a labyrinth without a map.”

The Apprentice's Confusion

A short distance away, a group of young apprentices from the merchant guild struggled with their abacuses. They attempted to tally their master's sales for the day, their brows furrowed in concentration. Harpreet approached them, offering her assistance. "It's tricky, isn't it?" she said kindly. "Imagine if we had a symbol for nothing. It would make this so much easier."

Amandeep, a bright-eyed apprentice, looked up at her. "But how can nothing be something? How can we have a number for what doesn't exist?"

Explaining the Concept

Harpreet smiled. "That's a good question, Amandeep. It's not about having a number for nothing, but a way to represent an empty space, a placeholder. It helps us keep things in order."

She took an abacus and showed them a series of beads representing different numbers. "Imagine you have five beads here and five beads here," she explained, moving the beads on either side of an empty space. "If you take away five from each side, you're left with nothing. But without a zero to mark that space, it's hard to see the balance."

LIRA'S CURIOSIT

Lira Kaur, who often wandered through the marketplace absorbing the sights and sounds, overheard the conversation. Her insatiable thirst for knowledge was piqued. "Baljit Uncle, Harpreet Aunty, how do you manage without zero?" she asked, her eyes wide with curiosity.

"It's a constant struggle, Lira," Baljit replied, shaking his head. "Balancing books, understanding profits and losses - everything is confusing without a way to represent nothing."

THE SCHOLAR'S INSIGHT

Rajinder Singh, a scholar from the Algebrion Institute, happened to be passing by and overheard their discussion. "You know, this issue isn't confined to commerce," he interjected. "In science, too, we face enormous challenges. Our calculations are often incomplete, leading to flawed conclusions."

"Exactly," Harpreet added. "Our entire system is built on approximations. We're constantly juggling numbers, trying to make sense of things."

THE COLLECTIVE REALIZATION

The group fell silent, each lost in their own thoughts. The absence of zero affected everyone in Numerica, from merchants and artisans to scholars and timekeepers. They had all learned to cope with its absence, but the inefficiencies were clear.

Baljit looked at Lira with a new sense of determination. "Maybe it's time we reconsider this ancient symbol," he said. "Perhaps zero isn't something to be feared but embraced."

SEEDS OF CHANGE

Lira's mind raced with possibilities. She had seen firsthand the impact of zero's absence on daily life. Baljit's words resonated with her, sparking a sense of purpose. She knew she had to delve deeper into this mystery, to uncover the true potential of zero.

With renewed determination, Lira began to formulate a plan. She would seek out forgotten knowledge, speak with more scholars, and gather evidence to convince the people of Numerica of zero's value. The journey ahead was daunting, but she felt ready to take it on.

A TURNING POINT

As the marketplace hummed with activity, Lira felt a shift in the air. The conversations she had overheard and participated in were the beginning of a change. The citizens of Numerica were starting to see the cracks in their system, the inefficiencies that plagued their daily lives.

With the support of Baljit, Harpreet, and Rajinder, Lira knew she could make a difference. The absence of zero had held Numerica back for too long. It was time to embrace the unknown, to challenge the status quo, and to unlock the city's true potential.

Chapter 3

III. THE ARCHITECT'S DILEMMA

The Renowned Architect

In Numerica, the name Amritjeet Kaur was synonymous with architectural brilliance. Her designs were admired for their beauty, though they were often plagued by structural issues. Amritjeet's work adorned the city with elegant towers and grand bridges, yet without zero, her precision in measurements always fell short, leading to unstable constructions.

The Importance of Precision

One day, as Amritjeet supervised the construction of a new amphitheater, a section of the structure began to crumble. "Stop! Everyone, move back!" she shouted, her heart sinking as she watched her work falter.

Lira Kaur, who had come to admire the construction, ran to Amritjeet's side. "What happened?" she asked, concern etched on her face.

"It's the same problem," Amritjeet replied, frustration evident in her voice. "Without zero, my measurements lack precision. We can't account for the smallest units, which means our calculations are always off."

THE FOUNDATION OF ZERO

Amritjeet led Lira to her workshop, a space filled with blueprints and models. She pointed to a scale drawing of the amphitheater. "Look here, Lira," she said, indicating the measurements. "Each line, each angle - they all depend on precise calculations. Without a way to represent nothing, to create a true starting point, everything is flawed."

Lira studied the drawings, seeing the gaps and inconsistencies that Amritjeet had to contend with. "So zero is like a foundation," Lira mused. "Without it, everything is unstable."

A CONVERSATION WITH A BUILDER

Nearby, Amritjeet's chief builder, Gurmeet Singh, was discussing the day's setback with his team. Gurmeet had worked with Amritjeet for years and understood the challenges they faced. "We need a way to make our work more reliable," he said. "Every collapse sets us back weeks, sometimes months."

He noticed Lira and Amritjeet approaching. "Amritjeet, do you think we'll ever solve this?" he asked.

"We must, Gurmeet," she replied. "Our city's future depends on it."

THE SCHOLAR'S VISIT

Rajinder Singh, who had heard of the recent collapse, decided to visit Amritjeet's workshop. "Amritjeet, Lira, I believe we're facing a common problem," he said as he entered. "Zero could be the solution."

Amritjeet nodded. "But how do we convince the city? They fear what they don't understand."

TEACHING THE CONCEPT

Rajinder suggested a demonstration. "Let's show them. Let's build a small model using the concept of zero."

They gathered a group of apprentices, including Amandeep and Baljit's son, Sukhdeep, to assist with the model. Amritjeet explained the importance of precision, using zero as a placeholder to ensure accurate measurements.

THE MODEL'S SUCCESS

The model, a miniature amphitheater, took shape over several days. Each piece fit perfectly, thanks to the inclusion of zero in their calculations. On the day of the unveiling, the city's leaders and curious citizens gathered to witness the demonstration.

As Amritjeet and Rajinder presented the model, the crowd was initially skeptical. But as they explained how zero had allowed them to create a flawless structure, the skepticism turned to amazement. The model stood firm, a testament to the power of accurate measurements.

OVERCOMING RESISTANCE

Not everyone was convinced. Some citizens, like the influential trader Ranjeet Singh, voiced their concerns. "This is too radical," he said. "We've managed without zero for generations. Why change now?"

Lira stepped forward. "Because we can do better, Uncle Ranjeet. Imagine the possibilities if we embrace this change. Our city could reach new heights, literally and figuratively."

THE TURNING POINT

The demonstration marked a turning point. Slowly, the citizens of Numerica began to see the potential of zero. Amritjeet's designs improved, becoming both beautiful and stable. Her reputation soared, and more architects and builders adopted the new system.

Gurmeet, now a firm believer, spoke to his team. "We're not just building structures. We're building a future."

A UNITED EFFORT

With the city's support growing, Amritjeet and Rajinder continued to advocate for zero's adoption. They held workshops and public lectures, spreading their knowledge and encouraging others to explore the concept.

Lira's journey had inspired a movement. What began as a quest for understanding had united Numerica in a shared goal: to embrace zero and unlock the full potential of their civilization. The Architect's Dilemma had become a story of innovation, collaboration, and the courage to challenge the status quo.

PRACTICAL APPLICATIONS

As the understanding and acceptance of zero grew, Numerica's architects began to see improvements in their work. Amritjeet Kaur's new projects were not only aesthetically pleasing but also structurally sound. Her colleagues, like Sukhbir Singh and Jasleen Kaur, who initially resisted the change, began incorporating zero into their designs. They noticed that their buildings stood longer and required fewer repairs.

In one such project, a new library, the team utilized zero-based calculations to ensure even weight distribution across the structure. This precision allowed for larger windows and higher ceilings, making the space not only functional but also inviting.

THE BRIDGE OF UNITY

Inspired by the success of the library, Amritjeet proposed an ambitious project: a bridge over the river Algebrion that would connect the two halves of Numerica. This bridge would symbolize the unity brought by the acceptance of zero.

The planning and construction involved the whole community. Lira Kaur gathered stories and testimonials from various citizens about how zero had improved their lives. These stories were engraved onto the bridge, making it a living testament to Numerica's transformation.

ACKNOWLEDGING CONTRIBUTIONS

During the bridge's construction, Amritjeet took the opportunity to mentor young architects and builders. She emphasized the importance of precision and the role zero played in their calculations. "Each of you is a part of Numerica's future," she told them. "Your willingness to embrace new ideas and challenge old norms will ensure our city's prosperity."

Among her mentees was Amandeep, the bright-eyed apprentice who had questioned the concept of zero. Under Amritjeet's guidance, he designed a smaller pedestrian bridge that won acclaim for its elegance and stability.

COMMUNITY SUPPORT

The bridge project united Numerica's citizens. Merchants like Baljit Singh and Harpreet Kaur contributed funds, while scholars like Rajinder Singh offered their knowledge. Even skeptical figures like Ranjeet Singh eventually came around to support the project, seeing the benefits of zero in the construction process.

THE GRAND UNVEILING

After months of meticulous planning and hard work, the bridge was finally completed. The citizens of Numerica gathered for the grand unveiling. Amritjeet, Lira, and the other key contributors stood proudly at the center of the bridge, ready to cut the ceremonial ribbon.

As the ribbon was cut, the crowd erupted in cheers. The bridge stood tall and strong, a symbol of Numerica's progress and unity. It was a testament to what could be achieved when the city embraced new ideas and worked together towards a common goal.

THE LASTING IMPACT

The success of the bridge had a ripple effect throughout Numerica. Other architects began incorporating zero into their designs, leading to safer and more efficient structures. The city's infrastructure improved, and Numerica entered a new era of architectural brilliance.

Amritjeet Kaur's influence extended beyond her designs. She became a mentor and a role model for young architects, encouraging them to think creatively and embrace new concepts. Her legacy lived on in the many structures that now adorned Numerica, each one a testament to the power of zero.

REFLECTIONS

One evening, as Amritjeet and Lira stood on the bridge they had helped create, they reflected on their journey. "We did it, Lira," Amritjeet said, her voice filled with pride. "We showed the city the power of zero."

Lira nodded. "And we proved that even the smallest concepts can have the biggest impact."

CONTINUING THE JOURNEY

As Numerica continued to grow and evolve, the story of zero and its acceptance became a foundational tale. It was taught in schools, discussed in public forums, and celebrated in annual festivals. The city had come to understand that embracing new ideas and challenging the status quo were essential for progress.

Amritjeet and Lira knew that their work was far from over. There were always new challenges to face and new ideas to explore. But they were confident that Numerica was on the right path, ready to face the future with an open mind and a willingness to embrace the unknown.

A LEGACY OF INNOVATION

The Architect's Dilemma had transformed into a legacy of innovation. The city of Numerica had learned to appreciate the value of zero and the importance of precision in all aspects of life. The citizens were more united, more curious, and more willing to explore new possibilities.

Amritjeet Kaur's name became synonymous with architectural excellence, and her story inspired future generations to think creatively and challenge the boundaries of what was possible. The city of Numerica thrived, and the lessons learned from the acceptance of zero continued to shape its future.

THE JOURNEY CONTINUES

As Amritjeet and Lira looked out over the city from their vantage point on the bridge, they knew that their journey was just beginning. There were still countless mysteries to uncover and challenges to overcome. The city's acceptance of zero was a monumental achievement, but it also opened up new realms of possibilities that they were eager to explore.

EMBRACING NEW IDEAS

Inspired by their success, Amritjeet and Lira decided to embark on new projects that pushed the boundaries of what they had previously thought possible. Amritjeet envisioned a network of canals and aqueducts that could revolutionize Numerica's water supply system. Lira, on the other hand, began working on educational programs to ensure that future generations understood the importance of zero and other mathematical concepts.

THE NETWORK OF CANALS

The construction of the canals was a massive undertaking that required the collaboration of engineers, builders, and mathematicians. Amritjeet's designs utilized zero-based calculations to ensure the even flow of water throughout the city. The project took years to complete, but when it was finished, Numerica's water supply was more reliable and efficient than ever before.

EDUCATIONAL PROGRAMS

Lira's educational programs were equally transformative. She worked with scholars and teachers to develop a curriculum that emphasized the importance of mathematical concepts, including zero. The programs were implemented in schools throughout Numerica, ensuring that every child had access to this essential knowledge. Lira's efforts helped create a new generation of thinkers who were equipped to build on the foundation she and Amritjeet had laid.

Acknowledging the Past

As the city moved forward, it also took time to honor its past. Monuments and plaques were erected to commemorate the journey of discovery that led to the acceptance of zero. The bridge over the river Algebrion became a symbol of unity and progress, drawing visitors from far and wide who came to learn about Numerica's remarkable transformation.

The Influence of Zero

The influence of zero extended beyond architecture and education. Merchants used it to improve their accounting methods, ensuring that their records were accurate and their businesses could grow. Scientists and astronomers found that zero allowed them to make more precise measurements and calculations, leading to new discoveries about the natural world.

The Spread of Knowledge

Word of Numerica's success spread to neighboring cities and regions. Delegations came to study Numerica's methods and learn from its scholars. Amritjeet and Lira welcomed these visitors, sharing their knowledge and encouraging other cities to embrace the power of zero. The ripple effect of their work began to influence the entire region, leading to a period of unprecedented growth and innovation.

Personal Growth

Through their work, both Amritjeet and Lira experienced personal growth. Amritjeet's confidence in her abilities as an architect soared, and she became a respected leader and mentor in her field. Lira, once a curious young girl, matured into a wise and compassionate educator, dedicated to nurturing the minds of future generations.

The Value of Collaboration

One of the most important lessons learned from their journey was the value of collaboration. The success of their projects relied on the combined efforts of many individuals, each contributing their unique skills and perspectives. Amritjeet and Lira realized that by working together, they could achieve far more than they ever could alone.

LOOKING TO THE FUTURE

As they stood on the bridge, Amritjeet and Lira looked out at the city with a sense of pride and hope. They knew that the journey of discovery was ongoing and that there would always be new challenges to face. But they were confident that Numerica was on a path to a brighter future, guided by the lessons they had learned and the values they had embraced.

A LEGACY OF CURIOSITY AND COURAGE

The story of Numerica and the acceptance of zero became a lasting legacy of curiosity and courage. It reminded everyone that progress often requires challenging the status quo and embracing new ideas. Amritjeet and Lira's journey had shown that even the smallest symbols could hold immense power and that the pursuit of knowledge was a noble and transformative endeavor.

As the sun set over the city, casting a golden glow on the bridge, Amritjeet and Lira knew that their work would inspire future generations to continue exploring, discovering, and building a better world. The Architect's Dilemma had become a story of triumph, innovation, and the unending quest for understanding.

CHAPTER

4

IV. THE SCHOLAR'S TALE

THE SCHOLAR'S DOMAIN

In the heart of Numerica, surrounded by towering shelves of ancient scrolls and manuscripts, lived Thalos, Lira's wise and revered grandfather. Thalos had dedicated his life to the pursuit of knowledge, and his home was a sanctuary of learning. The scent of old parchment filled the air, and sunlight streamed through narrow windows, casting a warm glow on the scholarly haven.

LIRA'S DISCOVERY

One afternoon, while helping Thalos organize his extensive collection, Lira stumbled upon a particularly ancient, dusty manuscript. The cover was adorned with a strange symbol - a simple circle. Intrigued, she carefully brought the scroll to her grandfather.

"Grandfather, what is this?" Lira asked, her eyes wide with curiosity.

Thalos peered over his spectacles, his eyes widening in recognition. "Ah, my dear, that is zero. It was introduced by a wandering sage long ago but was banished due to fear of its

power."

THE BANISHMENT OF ZERO

Lira's curiosity was piqued. "Why was it banished?"

"Many feared it would disrupt our numeric system," Thalos explained. "Zero was seen as a dangerous concept, capable of challenging our understanding of mathematics and the universe itself. The leaders of Numerica decided it was safer to erase all traces of it, hoping to maintain order and stability."

"But it was lost to time," Thalos sighed, "and with it, the potential for great advancements."

A DANGEROUS CONCEPT

Lira pondered her grandfather's words. The idea that a simple symbol could hold such power and be deemed dangerous fascinated her. She was determined to learn more about zero and its potential implications for Numerica.

Thalos, seeing the determination in Lira's eyes, decided to share more about the history of zero and its initial introduction to Numerica. He retrieved another ancient manuscript, this one detailing the sage who brought zero to the city.

THE WANDERING SAGE

The sage, Thalos recounted, was a traveler from distant lands who had spent years studying various mathematical systems. He had seen the transformative power of zero in other civilizations and believed that Numerica could greatly benefit from its adoption.

The sage's teachings initially intrigued the scholars of Numerica. They saw the potential for zero to simplify calculations and unlock new realms of scientific inquiry. However, as the implications of zero became clearer, fear and resistance grew.

FEAR AND RESISTANCE

"Many feared that zero would challenge the foundations of their beliefs," Thalos explained. "They worried it would disrupt the balance of their numerical systems and lead to chaos. The sage's ideas were met with increasing hostility."

The leaders of Numerica, influenced by these fears, decided to banish the concept of zero. The sage was forced to leave, and all records of zero were ordered to be destroyed. The manuscripts Lira had discovered were some of the few surviving remnants of this turbulent time.

THALOS'S HIDDEN KNOWLEDGE

Thalos confessed that he had secretly preserved these manuscripts, knowing that one day, Numerica might be ready to embrace zero again. He had always believed in the power of knowledge and the importance of understanding even the most controversial ideas.

Lira was moved by her grandfather's dedication to preserving knowledge. She realized that uncovering the truth about zero could be the key to solving many of Numerica's current problems. She was determined to continue her research and share her findings with the city.

THE IMPORTANCE OF ZERO

Thalos and Lira spent many evenings discussing the potential applications of zero. Thalos explained how zero could simplify complex calculations, provide a foundation for advanced mathematics, and offer new perspectives on the nature of existence.

"Zero is not just a number," Thalos said. "It represents the concept of emptiness, a bridge between the positive and the negative. It is a point of balance in the infinite dance of numbers."

Lira was fascinated by these ideas and began to see the profound impact zero could have on various aspects of life in Numerica. She felt a growing sense of responsibility to share this knowledge with others.

PREPARING FOR THE REVELATION

With Thalos's guidance, Lira prepared to share her discoveries with the people of Numerica. They gathered evidence, created demonstrations, and developed a clear and compelling explanation of zero's benefits.

Thalos encouraged Lira to be patient and understanding, knowing that many people would be skeptical and resistant to change. He emphasized the importance of education and communication in overcoming fear and ignorance.

THE SCHOLAR'S TALE UNFOLDS

As Lira prepared for her grand presentation, she reflected on the journey that had brought her to this point. She realized that her quest for knowledge was not just about discovering zero but about understanding the deeper truths of the universe and sharing them with others.

The Scholar's Tale was a story of courage, curiosity, and the enduring power of knowledge. Lira knew that her journey was far from over, but she felt ready to face whatever challenges lay ahead. She was determined to honor her grandfather's legacy and help Numerica embrace the transformative power of zero. The more she delved into the manuscripts, the more she understood the profound implications of this simple concept. She envisioned a Numerica where calculations were precise, architecture was stable, and time was accurately measured. But to make this vision a reality, she needed the city's acceptance and support.

THE GRAND PRESENTATION

The day of the presentation arrived. The grand amphitheater was filled with scholars, merchants, architects, timekeepers, and curious citizens. Lira stood at the center, her heart pounding with a mix of excitement and nervousness. Thalos watched proudly from the front row, his eyes twinkling with encouragement.

"Ladies and gentlemen of Numerica," Lira began, her voice steady and clear, "I am here to share with you a concept that has the power to transform our city. A concept that has been forgotten and feared, but which holds immense potential - the concept of zero."

Demonstrating Zero's Power

Lira unveiled the ancient manuscripts, showing the simple circle that represented zero. She explained how zero could simplify complex calculations, provide a foundation for advanced mathematics, and offer new perspectives on the nature of existence.

To demonstrate its practical applications, Lira used examples from everyday life. She showed how zero could help merchants balance their books, architects create stable designs, and timekeepers accurately track the passage of time. Her explanations were clear and compelling, and she could see the spark of understanding in the eyes of her audience.

Addressing Skepticism

As expected, there were skeptics in the crowd. Some questioned the necessity of zero, while others feared its potential to disrupt their lives. Lira listened patiently to their concerns, addressing each one with thoughtful and reasoned responses.

"Change can be daunting," she acknowledged, "but it is through embracing new ideas that we grow and evolve. Zero is not something to fear, but a tool to help us achieve greater precision, understanding, and harmony in our lives."

Community Support

Gradually, the atmosphere in the amphitheater shifted. Merchants like Baljit Singh and Harpreet Kaur began to see the benefits of zero for their accounting. Architects like Amritjeet Kaur imagined the stability it could bring to their designs. Even skeptical figures like Ranjeet Singh started to reconsider their positions.

One by one, members of the community stood up to voice their support. The bridge project had already shown them the power of collaboration, and they were willing to give zero a chance.

Thalos's Encouragement

Throughout the presentation, Thalos's presence provided Lira with unwavering support. Afterward, he approached her with a proud smile.

"You did wonderfully, Lira," he said, embracing her. "You have the heart of a scholar and the courage of a leader. Numerica is ready to embrace zero, thanks to you."

Lira felt a surge of gratitude and determination. She knew that the journey was far from over, but with the support of her grandfather and the community, she was ready to face whatever challenges lay ahead.

IMPLEMENTING ZERO

With the community's backing, Lira and Thalos began the process of integrating zero into Numerica's systems. They worked with merchants to update accounting methods, architects to revise their designs, and timekeepers to recalibrate their instruments.

The transition was not without its difficulties. There were setbacks and resistance, but Lira's patience and persistence paid off. Gradually, zero became an accepted and valued part of Numerica's mathematical landscape.

EDUCATIONAL REFORMS

Recognizing the importance of education, Lira and Thalos developed a curriculum that emphasized the significance of zero. Schools throughout Numerica adopted the new program, ensuring that future generations would understand and appreciate this fundamental concept.

Lira took a hands-on approach, visiting classrooms and engaging with students. She shared the story of zero's journey and the lessons it held about curiosity, courage, and the power of knowledge.

THE RIPPLE EFFECT

The impact of zero extended beyond Numerica. News of the city's transformation spread to neighboring regions, attracting scholars and leaders who were eager to learn from Lira and Thalos's experiences.

Delegations visited Numerica to study its methods and incorporate zero into their own systems. Lira welcomed these visitors, sharing her knowledge and encouraging a spirit of collaboration and innovation.

A NEW ERA OF ENLIGHTENMENT

As Numerica fully embraced zero, the city entered a new era of enlightenment and prosperity. The precision and clarity that zero brought to mathematics and daily life led to unprecedented advancements in science, architecture, commerce, and timekeeping.

Lira and Thalos continued their work, always seeking new ways to expand their understanding and share their knowledge. Their legacy of curiosity and courage inspired others to explore the unknown and push the boundaries of human

A NEW ERA OF ENLIGHTENMENT

Understanding. Numerica flourished, becoming a beacon of progress and innovation in the ancient world. The city was no longer just a hub of trade and culture; it was now a center of learning and discovery, where the pursuit of knowledge was valued above all else.

LIRA'S CONTINUED JOURNEY

As the years passed, Lira's dedication to education and innovation never waned. She continued to explore the depths of mathematics, driven by an insatiable curiosity. Her reputation as a scholar and teacher grew, and she was often invited to speak at gatherings and conferences beyond Numerica's borders.

One day, while reflecting on her journey, Lira realized that her quest for knowledge had only just begun. There were still many mysteries to uncover, many concepts to explore. She felt a renewed sense of purpose and determination to push the boundaries of what was known and to inspire others to do the same.

THE ENDURING LEGACY

Thalos, now elderly but still sharp of mind, often looked upon his granddaughter with pride. He knew that Lira's work would continue to shape Numerica for generations to come. Her story was a testament to the power of curiosity, courage, and the relentless pursuit of knowledge.

Together, they had transformed a city, bridging the gap between the past and the future with the simple yet profound concept of zero. Their legacy was not just in the numbers and equations, but in the spirit of inquiry and discovery that they had instilled in the people of Numerica.

Chapter

5

V. THE TIMEKEEPER'S STRUGGLE

The Master of Time

In the heart of Numerica, nestled among the bustling streets and grand structures, stood the Clock Tower, a symbol of the city's attempt to grasp the flow of time. At the helm of this effort was Eldrin, the city's dedicated timekeeper. Eldrin's life revolved around the complex gears and mechanisms of the Clock Tower, ensuring that the city's schedules and events ran smoothly.

The Challenge of Accuracy

Eldrin faced immense challenges in his role. Without zero, tracking the passage of time accurately was a near-impossible task. The absence of a starting point made it difficult to calculate durations, and as a result, the city's timekeeping was often unreliable.

One evening, while adjusting the intricate gears of the Clock

Tower, Eldrin confided in Lira, who had become a frequent visitor. "Imagine trying to schedule events or calculate durations without a starting point," he told her. "Zero is crucial for marking the beginning of cycles. Without it, our calendars and clocks are always off."

Lira's Growing Understanding

Lira saw the frustration in Eldrin's eyes and realized the profound impact zero's absence had on their lives. She had already learned about zero's importance in commerce and architecture, and now she was beginning to understand its critical role in timekeeping.

"Eldrin, what if we could find a way to introduce zero into our timekeeping system?" Lira suggested. "It could bring the precision and accuracy we desperately need."

Eldrin looked thoughtful. "It's a radical idea, Lira. But if anyone can find a way, it's you."

Exploring Ancient Timekeeping

Determined to help Eldrin and the city, Lira delved into ancient manuscripts and texts, searching for clues about timekeeping methods that might incorporate zero. She discovered that other civilizations had used zero as a placeholder, allowing for more precise calculations of time.

Lira shared her findings with Eldrin, who was both fascinated and skeptical. "These methods seem promising, but implementing them will be a massive undertaking. We'll need the support of the entire city."

The First Experiment

Undeterred, Lira and Eldrin decided to conduct an experiment. They created a small, experimental clock that incorporated the concept of zero. The clock was rudimentary, but it demonstrated the potential for increased accuracy.

They presented their prototype to the city's council, explaining how zero could revolutionize timekeeping in Numerica. The council members were intrigued but cautious, aware of the challenges that lay ahead.

BUILDING SUPPORT

Lira and Eldrin knew that gaining widespread support was crucial. They began a campaign to educate the citizens about the benefits of zero in timekeeping. They held public demonstrations, showing how the experimental clock worked and explaining the concept in simple terms.

Merchants, who had already seen the benefits of zero in their accounting, were among the first to support the initiative. They recognized that accurate timekeeping could improve their business operations and help them coordinate with customers and suppliers more effectively.

OVERCOMING RESISTANCE

Not everyone was convinced. Some citizens were resistant to change, fearing that the new timekeeping system would disrupt their lives. Lira and Eldrin faced questions and doubts at every turn, but they remained patient and persistent.

One particularly vocal opponent was Arjun Singh, a traditionalist who believed that the old ways were best. "This newfangled idea of zero is unnecessary," he argued. "We've managed without it for centuries. Why change now?"

Lira addressed his concerns with respect and clarity. "Arjun Singh, the world is constantly evolving, and so must we. Zero is not just a number; it's a tool that can help us understand and navigate the complexities of time. By embracing it, we can enhance our lives and ensure a better future for Numerica."

THE TURNING POINT

As Lira and Eldrin continued their efforts, more and more citizens began to see the potential benefits of zero in timekeeping. The turning point came when a particularly important city event, the annual Festival of Lights, was nearly ruined due to inaccurate timekeeping.

Eldrin's experimental clock saved the day, allowing the festival to proceed smoothly. This success demonstrated the practical advantages of incorporating zero and won over many skeptics.

IMPLEMENTING THE NEW SYSTEM

With growing support, the city council approved the implementation of the new timekeeping system. Eldrin and Lira worked tirelessly to upgrade the Clock Tower and integrate zero into its mechanisms. They also trained a team of apprentices, ensuring that the knowledge and skills needed to maintain the system would be passed down to future generations.

The transition was challenging, but the benefits quickly became apparent. Schedules became more reliable, events were better coordinated, and the city's daily life became more orderly and efficient.

THE TIMEKEEPER'S TRIUMPH

Eldrin's struggles as a timekeeper were transformed into triumphs. The Clock Tower, once a symbol of frustration and inaccuracy, became a testament to the power of innovation and the importance of zero. Eldrin, once burdened by the challenges of his role, now found joy and satisfaction in his work.

Lira's efforts had once again brought about significant change in Numerica. She had helped the city embrace zero in yet another vital aspect of life, proving that even the most daunting challenges could be overcome with knowledge, determination, and collaboration.

A NEW CHAPTER

As Eldrin and Lira stood together, looking up at the Clock Tower, they felt a deep sense of accomplishment. Their journey was far from over, but they had made great strides in helping Numerica progress and thrive.

Lira knew that there were still many more challenges to face and discoveries to make. But with the support of her friends and the lessons she had learned along the way, she was ready to continue her quest for knowledge and innovation.

THE LEGACY OF TIME

The transformation of Numerica's timekeeping system became a lasting legacy. The city's newfound precision and efficiency paved the way for further advancements in science, technology, and daily life. Eldrin and Lira's names were remembered as pioneers who had dared to embrace change and unlock the potential of zero.

As Numerica continued to grow and evolve, the lessons of the past served as a guiding light. The citizens understood the importance of curiosity, courage, and collaboration in shaping a better future. And in the heart of the city, the Clock Tower stood as a timeless reminder of the power of zero and the journey that had brought them to this point.

BRIDGING THE PAST AND FUTURE

Eldrin and Lira worked together to create a comprehensive training program for future timekeepers. They meticulously documented their discoveries, ensuring that the knowledge of zero and its applications in timekeeping would not be lost to future generations.

They also collaborated with other scholars and innovators in Numerica, sharing their findings and encouraging interdisciplinary studies. This collaborative spirit fostered a sense of unity and progress within the city, further solidifying Numerica's reputation as a hub of learning and innovation.

THE EVOLUTION OF TIMEKEEPING

As the new timekeeping system was fully integrated into Numerica's daily life, the city experienced a renaissance of punctuality and efficiency. Public transportation ran smoothly, events started and ended on time, and businesses thrived with better coordination.

The success of the timekeeping system also inspired other innovations. Eldrin and Lira's work demonstrated the power of mathematical concepts to solve real-world problems, sparking interest in further exploration of numerical theory and its practical applications.

ELDRIN'S PERSONAL GROWTH

Through his collaboration with Lira, Eldrin experienced significant personal growth. He had always been dedicated to his work, but the introduction of zero had transformed his perspective. He now understood the importance of flexibility, open-mindedness, and the willingness to embrace new ideas.

Eldrin's newfound confidence and sense of purpose resonated with his apprentices and colleagues. He became a mentor and a role model, inspiring others to pursue their passions and to seek innovative solutions to the challenges they faced.

LIRA'S CONTINUED QUEST

Lira's journey did not end with the transformation of Numerica's timekeeping system. She continued to seek out ancient manuscripts and texts, always eager to learn more about the mathematical concepts that underpinned their world.

Her travels took her to distant lands and brought her into contact with other scholars and thinkers. Through these interactions, Lira expanded her understanding of mathematics and its applications, continually refining her knowledge and bringing new insights back to Numerica.

THE IMPACT OF ZERO ON SOCIETY

As Numerica fully embraced the concept of zero, the impact on society was profound. The city's infrastructure improved, public services became more efficient, and the overall quality of life increased.

The citizens of Numerica grew more open to innovation and progress. They began to view challenges as opportunities for growth and discovery, fostering a culture of continuous improvement and learning.

THE POWER OF KNOWLEDGE

Lira's dedication to the pursuit of knowledge inspired many in Numerica. Her story became a symbol of the transformative power of education and the importance of questioning established norms.

Schools and academies flourished, with an emphasis on critical thinking, problem-solving, and the application of mathematical principles. The city's youth were encouraged to explore new ideas and to contribute to the ongoing progress of Numerica.

A NEW ERA OF ENLIGHTENMENT

Numerica entered a new era of enlightenment, characterized by a deep appreciation for the power of knowledge and the potential of the human mind. The city's advancements in mathematics, science, and technology attracted visitors and scholars from around the world.

Numerica became a beacon of progress and innovation, a place where ideas were valued and nurtured. The city's success demonstrated the importance of embracing change and the potential of even the simplest concepts to bring about profound transformation.

LIRA'S CONTINUED JOURNEY

As the years passed, Lira's dedication to learning and discovery never wavered. She continued to travel far and wide, seeking out hidden knowledge and forgotten wisdom. Her reputation as a brilliant scholar and innovator grew, and she was often invited to share her insights with other cities and cultures.

On one of her journeys, Lira encountered a group of mathematicians who were exploring the concept of negative numbers. Intrigued, she delved into their research and realized how zero served as the critical pivot point between positive and negative values. This realization deepened her understanding of zero's fundamental role in the broader mathematical framework.

COLLABORATIVE BREAKTHROUGHS

Lira's collaboration with these mathematicians led to several groundbreaking discoveries. Together, they developed new mathematical models and theories that expanded the boundaries of known science. These advancements had practical applications in various fields, from engineering to astronomy.

Upon her return to Numerica, Lira shared these discoveries with the city's scholars and innovators. The collaborative spirit that she had fostered continued to thrive, leading to a series of innovations that further propelled Numerica into a new era of progress and enlightenment.

THE SYMBOL OF UNITY

The story of zero and its impact on Numerica became a powerful symbol of unity and collaboration. The city's transformation served as a testament to the idea that even the most challenging problems could be overcome through collective effort and shared knowledge.

Public monuments and artworks were created to celebrate the introduction of zero, depicting the journey of discovery and the profound changes it brought about. These symbols served as a reminder of the city's commitment to innovation and progress.

PERSONAL GROWTH AND REFLECTION

Lira's journey also brought about significant personal growth. She had started as a curious young girl eager to learn, and she had evolved into a wise and respected leader in the field of mathematics. Her experiences taught her the value of perseverance, open-mindedness, and the courage to challenge established norms.

Reflecting on her journey, Lira realized that her greatest achievements were not just the discoveries she had made but the impact she had on the lives of others. Her efforts had inspired a new generation of thinkers and innovators, ensuring that Numerica's legacy of progress would continue for years to come.

A LEGACY OF KNOWLEDGE

As Lira grew older, she focused on preserving her knowledge and experiences for future generations. She wrote extensive treatises on her discoveries and the importance of zero, ensuring that the lessons she had learned would not be forgotten.

She also established a foundation dedicated to promoting education and innovation in Numerica. This foundation provided scholarships and resources for young scholars, encouraging them to explore new ideas and contribute to the ongoing advancement of society.

The Wisdom of Elders

Lira's relationship with her grandfather, Thalos, remained a source of strength and inspiration throughout her life. Thalos had passed away, but his teachings and wisdom continued to guide her. She often reflected on his words and the lessons he had imparted, finding comfort and direction in his legacy.

In honor of her grandfather, Lira established a grand library in Numerica, filled with ancient manuscripts and modern texts. This library became a center of learning and exploration, attracting scholars from all corners of the world.

Bridging Generations

Lira's efforts to bridge the past and the future created a lasting impact on Numerica. The city's citizens understood the importance of honoring their heritage while embracing new ideas and innovations. This balance ensured that Numerica remained a vibrant and dynamic place, continuously evolving and improving.

Young and old alike were inspired by Lira's story. Her journey served as a reminder that knowledge was a lifelong pursuit and that the quest for understanding could bring about profound change.

A NEW DAWN

As Numerica continued to flourish, Lira felt a deep sense of fulfillment. She had dedicated her life to the pursuit of knowledge and the betterment of her city, and she had succeeded beyond her wildest dreams.

One morning, as she stood atop the Clock Tower, Lira looked out over Numerica and saw a city transformed. The streets were bustling with activity, the architecture was more magnificent than ever, and the city's timekeeping was precise and reliable. She knew that her journey had been worthwhile and that the impact of her discoveries would be felt for generations to come.

THE CELEBRATION OF ZERO

In honor of Lira's contributions, the city of Numerica declared an annual celebration called "Zero Day." This event commemorated the introduction of zero and its profound impact on the city's development. Citizens gathered to participate in lectures, workshops, and festivities that celebrated the power of knowledge and innovation.

Lira was often the guest of honor at these celebrations, sharing her experiences and encouraging others to pursue their own quests for understanding. Her presence was a source of inspiration, reminding everyone of the importance of curiosity and the courage to explore the unknown.

PASSING THE TORCH

As Lira approached the twilight of her life, she focused on mentoring the next generation of scholars and innovators. She passed on her knowledge, guiding young minds and helping them navigate the complexities of mathematics and science.

Lira knew that her journey would continue through the work of those she had inspired. She had planted the seeds of curiosity and innovation, and she watched with pride as they grew and flourished.

A LEGACY ETCHED IN TIME

Lira's legacy was etched in the very fabric of Numerica. Her contributions had transformed the city and set it on a path of continuous progress and enlightenment. The story of zero, once a forgotten concept, had become a symbol of hope and possibility.

As Lira looked out over the city one last time, she felt a profound sense of peace. She knew that Numerica was in good hands and that the journey of discovery would continue. With a heart full of gratitude and a mind brimming with memories, Lira embraced the next chapter of her existence, confident that the lessons of zero would guide her city for generations to come.

THE ENDURING JOURNEY

Even after Lira's passing, her influence persisted. The principles she had championed continued to drive Numerica's progress. Her teachings were embedded in the city's educational system, ensuring that the spirit of curiosity and innovation lived on.

Numerica's citizens, inspired by Lira's legacy, continued to explore new frontiers of knowledge. They understood that the journey of discovery was never truly complete and that each generation had the responsibility to build upon the foundations laid by those who came before.

A TRIBUTE TO LIRA

In recognition of Lira's immeasurable contributions, Numerica erected a statue in her honor. The statue depicted Lira holding a book and a clock, symbolizing her dedication to knowledge and timekeeping. It stood in the city center, a reminder of the power of a single individual's quest for understanding.

Visitors from around the world came to Numerica to pay tribute to Lira and to learn from her story. Her legacy had transcended the boundaries of her city, inspiring people everywhere to embrace the unknown and to seek the truths hidden within the fabric of the universe.

THE FUTURE UNFOLDS

As Numerica moved forward, it faced new challenges and opportunities. The city's commitment to knowledge and innovation remained unwavering, guided by the lessons of the past and the promise of the future.

Lira's journey had shown that even the smallest concepts could hold immense power. Zero, once a simple circle on an ancient manuscript, had transformed Numerica and its people. The city's progress was a testament to the enduring power of curiosity, collaboration, and the courage to explore the unknown.

And so, the story of Numerica and its journey with zero continued, a never-ending quest for knowledge and enlightenment. The legacy of Lira and the lessons of zero would forever guide the city's path, illuminating the way forward and inspiring generations to co

CHAPTER 6

VI. LIRA'S QUEST

THE CALL TO ADVENTURE

Determined to uncover the secrets of zero, Lira embarked on a journey to the farthest reaches of Numerica. Her quest was driven by an insatiable thirst for knowledge and a desire to understand the profound mysteries of mathematics. She prepared for her journey meticulously, gathering supplies, maps, and books.

Her departure was met with both excitement and apprehension. Friends and family wished her well, though they worried about the challenges she might face. Lira reassured them with her characteristic optimism and unwavering resolve, promising to return with the knowledge that could change their lives.

THE FORGOTTEN LIBRARIES

Lira's first destination was the ancient Library of Arithmos, a place long believed to hold forgotten wisdom. The library, hidden deep within a dense forest, was a marvel of ancient architecture, with towering shelves filled with dusty scrolls and manuscripts.

As Lira delved into the library's depths, she encountered texts that hinted at the existence of zero. These manuscripts, written by scholars from centuries past, described the concept as both revolutionary and dangerous. They spoke of a time when zero was known but later suppressed due to fear and misunderstanding.

HIDDEN ALCOVES

Continuing her journey, Lira discovered hidden alcoves where ancient scholars once conducted their research. These secluded spots, often in remote and rugged terrain, provided her with more clues about the significance of zero. She found inscriptions and diagrams that illustrated how zero could transform calculations and measurements.

In one such alcove, Lira stumbled upon a hidden chamber containing an ancient artifact - a stone tablet inscribed with mathematical symbols and equations. Among these was the elusive symbol of zero, depicted as a simple, elegant circle. This discovery filled her with awe and renewed determination.

MEETING FELLOW SEEKERS

Throughout her travels, Lira met other seekers of knowledge. Scholars, merchants, architects, and timekeepers all shared their struggles and experiences with her. They were intrigued by her quest and offered their insights and assistance.

One such scholar, Dr. Amandeep Kaur, had been researching the history of zero for years. She provided Lira with invaluable information about how zero was once an integral part of Numerica's mathematical system before it was erased from common knowledge. Together, they pieced together the puzzle of zero's disappearance and its potential to revolutionize their world.

THE TRIALS OF THE QUEST

Lira's journey was fraught with challenges. She faced harsh weather, treacherous landscapes, and the constant threat of failure. But her determination never wavered. Each trial only strengthened her resolve to uncover the truth about zero.

During one particularly arduous climb up a mountain to reach an ancient observatory, Lira encountered harsh storms and treacherous paths. However, her perseverance paid off when she discovered an ancient manuscript detailing how zero was used in astronomical calculations to predict celestial events with remarkable accuracy.

A REVELATION IN THE DESERT

One of the most significant discoveries came during Lira's journey through the desert of Equatia. She arrived at the ruins of an ancient city, where she found a hidden library beneath the sands. The scrolls and tablets there revealed how zero had once been used to simplify complex equations and understand the universe's deeper truths.

In these manuscripts, Lira found detailed explanations of zero's role as a bridge between positive and negative numbers, a concept that revolutionized her understanding of mathematics. This discovery marked a turning point in her quest, solidifying her belief in the power of zero.

GATHERING ALLIES

As Lira traveled back to Numerica, she gathered a group of allies who shared her vision. These included mathematicians, engineers, and other scholars who recognized the potential of zero. Together, they formed a coalition dedicated to reintroducing zero into Numerica's mathematical framework.

One of her most influential allies was Professor Jaspreet Singh, a renowned mathematician who had long advocated for the use of zero. His expertise and reputation helped lend credibility to their cause, making it easier for them to gain support from other scholars and officials.

PREPARING FOR THE REVELATION

Upon returning to Numerica, Lira and her allies prepared for a grand presentation. They knew that introducing zero to the public would be met with skepticism and resistance, so they planned a series of demonstrations to show zero's practical applications and benefits.

They worked tirelessly, creating models, conducting experiments, and preparing speeches. Lira's home became a hub of activity as scholars and innovators collaborated to perfect their presentation. The excitement was palpable, and Lira felt a sense of anticipation and hope.

THE GRAND GATHERING

The day of the grand gathering arrived. The citizens of Numerica filled the city's grand amphitheater, curious to hear what Lira and her allies had to say. The atmosphere was charged with anticipation as Lira took the stage, her heart pounding with a mixture of nervousness and excitement.

She began by sharing the story of her journey, describing the trials she faced and the discoveries she made. She spoke passionately about the importance of zero and how it could transform their understanding of mathematics and the world.

DEMONSTRATING THE POWER OF ZERO

Lira and her allies then demonstrated the practical applications of zero. They showed how it simplified complex calculations, brought precision to measurements, and offered new perspectives on the nature of existence. Merchants saw how zero could balance their accounts, architects realized how it could stabilize their designs, and timekeepers marveled at the accuracy it brought to their work.

The demonstrations were met with a mixture of amazement and skepticism. But as Lira continued to explain and illustrate the benefits of zero, the audience began to see its potential. Questions were asked, and Lira answered them with clarity and conviction, gradually winning over the crowd.

OVERCOMING RESISTANCE

Despite the initial excitement, there was still resistance. Some citizens clung to their old ways, fearing the changes that zero would bring. They questioned its validity and usefulness, wary of disrupting the established order.

Lira addressed these concerns with patience and understanding. She acknowledged their fears but emphasized that embracing zero would lead to greater prosperity and enlightenment. She drew parallels to other innovations that had initially faced resistance but ultimately improved their lives.

A UNIFIED VISION

Gradually, the tide began to turn. The practical demonstrations, combined with Lira's passionate advocacy, convinced more and more citizens of zero's value. Merchants, architects, and timekeepers who had been skeptical started to see the benefits in their daily lives.

The city's leaders, recognizing the growing support for zero, decided to endorse its reintroduction formally. They saw the potential for progress and innovation that zero could bring and understood the importance of embracing change.

CELEBRATING A NEW ERA

With the city's leaders on board, Numerica officially embraced zero. The grand amphitheater erupted in applause and celebration as the citizens recognized the significance of this moment. The reintroduction of zero marked the beginning of a new era of prosperity and enlightenment.

Festivities were held throughout the city, celebrating the return of zero and the possibilities it brought. Lira and her allies were honored for their dedication and perseverance, hailed as heroes who had transformed Numerica.

LIRA'S CONTINUED JOURNEY

Though Lira had achieved her goal, her journey was far from over. She continued to explore new frontiers of knowledge, driven by an unending curiosity and a desire to understand the universe's deeper truths.

Her work inspired others to pursue their quests for knowledge, creating a ripple effect that spread throughout Numerica and beyond. The city's commitment to innovation and progress remained strong, guided by the principles of curiosity and collaboration.

A LEGACY OF KNOWLEDGE

Lira's legacy lived on in the hearts and minds of Numerica's citizens. Her journey had shown them the power of a single idea and the importance of embracing the unknown. Zero, once a forgotten concept, had become a symbol of hope and possibility, inspiring future generations to explore the mysteries of the universe.

The story of Lira's quest and the reintroduction of zero was etched into Numerica's history, serving as a reminder of the enduring power of knowledge and the courage to seek the truth. The city's journey with zero continued, a never-ending quest for understanding and enlightenment.

ESTABLISHING NEW INSTITUTIONS

With the acceptance of zero, Lira and her allies established new institutions to educate Numerica's citizens about its applications and benefits. They founded the Zero Institute, dedicated to advancing mathematical knowledge and fostering innovation. The institute attracted scholars from all over Numerica and beyond, eager to explore the new possibilities zero offered.

Classes, lectures, and workshops were held regularly, with Lira often at the helm, sharing her discoveries and insights. The institute quickly became a beacon of learning, its influence spreading across the region.

TRANSFORMING EDUCATION

The introduction of zero transformed Numerica's educational system. Schools adopted new curricula that incorporated zero into all aspects of learning. Students were taught its importance from an early age, and as a result, they grew up with a deep understanding of its value.

Teachers embraced the new methods, using zero to simplify complex concepts and make learning more engaging. The effects were profound, with students demonstrating improved problem-solving skills and a greater enthusiasm for mathematics and science.

ENGINEERING BREAKTHROUGHS

The field of engineering experienced a revolution. Architects like Amritjeet Singh used zero to create more precise and stable structures, pushing the boundaries of design. New buildings, bridges, and other infrastructures began to appear, showcasing innovative techniques that were previously unimaginable.

One of the most notable projects was the construction of a new aqueduct system that brought fresh water from the mountains to Numerica. The project, led by Harpreet Kaur, utilized zero-based calculations to ensure efficiency and reliability. It became a model of modern engineering, admired by neighboring cities.

ECONOMIC GROWTH

The reintroduction of zero spurred economic growth. Merchants like Baljit Singh found their trade records more manageable and accurate, leading to better business practices. The ability to balance accounts and track profits with precision resulted in increased trade and prosperity.

Numerica's marketplace thrived, attracting traders from distant lands. The city became a hub of commerce, its economy flourishing as a result of the newfound efficiency and reliability that zero brought.

Advancements in Timekeeping

Timekeeper Eldrin's work saw significant improvements. With zero, the accuracy of calendars and clocks reached new heights. Eldrin and his colleagues developed more precise timekeeping devices, allowing citizens to schedule events and manage their time more effectively.

The Improved timekeeping methods had a ripple effect across various aspects of life, from agriculture to transportation. Farmers could better plan their planting and harvesting cycles, while travelers benefited from reliable schedules for their journeys.

Scientific Discoveries

The acceptance of zero opened up new avenues for scientific exploration. Scholars like Dr. Amandeep Kaur delved deeper into the mysteries of the universe, using zero to simplify complex equations and model natural phenomena more accurately.

One of the most groundbreaking discoveries was in the field of astronomy. Using zero-based calculations, Numerican astronomers accurately predicted celestial events, enhancing their understanding of the cosmos. This led to a renewed interest in space exploration and the development of new observational techniques.

CULTURAL RENAISSANCE

The reintroduction of zero also sparked a cultural renaissance. Artists and musicians found inspiration in the concept of zero, incorporating it into their work in innovative ways. Paintings, sculptures, and compositions reflected the harmony and balance that zero represented.

Numerica's cultural scene flourished, with festivals and exhibitions celebrating the intersection of art and mathematics. The city became known not only for its intellectual achievements but also for its vibrant cultural life.

POLITICAL STABILITY

The acceptance of zero contributed to political stability. Leaders recognized the benefits it brought to governance, from accurate record-keeping to efficient resource management. Policies and decisions were based on reliable data, leading to more effective and transparent governance.

The citizens of Numerica, seeing the positive changes brought by zero, placed greater trust in their leaders. This trust fostered a sense of unity and cooperation, further strengthening the city's social fabric.

PERSONAL GROWTH

On a personal level, Lira's journey with zero transformed her life. She grew from a curious young girl into a respected scholar and leader, her experiences shaping her into a person of wisdom and compassion. Her dedication to knowledge and her willingness to face challenges inspired those around her.

Lira's relationship with her grandfather, Thalos, deepened as they shared their discoveries and worked together to promote the benefits of zero. Thalos took great pride in Lira's accomplishments, seeing in her the fulfillment of his lifelong quest for knowledge.

MENTORING THE NEXT GENERATION

As Lira continued her work, she took on the role of a mentor, guiding the next generation of scholars and innovators. She encouraged young minds to explore the unknown and to approach challenges with curiosity and determination.

One of her most promising students, a young mathematician named Simran Kaur, showed exceptional talent and enthusiasm. Under Lira's guidance, Simran developed new mathematical theories and contributed to significant advancements in the field.

EXPANDING HORIZONS

The success of zero in Numerica inspired other cities to follow suit. Scholars and leaders from neighboring regions visited Numerica to learn about its implementation and effects. The exchange of knowledge and ideas led to collaborative projects and shared advancements.

Numerica's influence spread far and wide, becoming a center of learning and innovation. The city's commitment to progress and its embrace of zero served as a model for others, fostering a spirit of cooperation and mutual growth.

PRESERVING KNOWLEDGE

To ensure that the knowledge of zero was preserved for future generations, Lira and her allies established the Great Archive of Numerica. This vast repository housed manuscripts, artifacts, and records related to zero and its applications.

The Great Archive became a place of pilgrimage for scholars and students, a testament to Numerica's dedication to the pursuit of knowledge. It served as a reminder of the city's journey and the importance of preserving and sharing wisdom.

HONORING TRADITION

While embracing zero brought many changes, Numerica also honored its traditions. The city's rich history and cultural heritage were celebrated alongside its modern advancements. Festivals, ceremonies, and rituals continued to play a vital role in the community's life.

Lira's efforts to integrate zero into Numerica's traditions ensured that the concept was not seen as a disruption but as an enhancement of their cultural identity. This balance between tradition and innovation became a defining characteristic of Numerica.

REFLECTING ON THE JOURNEY

As years passed, Lira often reflected on her journey. She thought about the challenges she faced, the discoveries she made, and the impact her work had on Numerica. Her quest for zero had transformed not only her life but the lives of countless others.

Lira's reflections were captured in her writings, which became treasured texts for future generations. Her journals and memoirs provided insight into her thoughts and experiences, inspiring others to embark on their quests for knowledge.

LOOKING TO THE FUTURE

Numerica's journey with zero was a story of transformation, progress, and enlightenment. As the city looked to the future, it remained committed to the values of curiosity, collaboration, and the pursuit of knowledge.

Lira's legacy lived on, a beacon of hope and inspiration. Her quest for zero had shown that even the smallest and most unassuming symbols could hold the greatest significance. Numerica's story was a testament to the power of curiosity and the courage to explore the unknown, a journey that would continue to inspire generations to come.

CHAPTER 7

VII. THE REVELATION

A GRAND GATHERING

With newfound understanding, Lira returned to Numerica, ready to share her discoveries. She called for a gathering at the city's grand amphitheater, where she unveiled the secrets of zero to the people. She demonstrated how it could simplify complex calculations, bring precision to measurements, and offer new perspectives on the nature of existence.

The amphitheater buzzed with anticipation as citizens from all walks of life assembled. Merchants, architects, scholars, and timekeepers filled the seats, their faces reflecting a mixture of curiosity and skepticism. Lira stood at the center, her heart pounding with a mix of excitement and nervousness.

THE INITIAL SKEPTICISM

"Friends, I have journeyed far and wide, seeking knowledge and understanding," Lira began. "Today, I bring you the concept of zero, a symbol that can change our lives in ways we have never imagined."

As Lira spoke, she could see the doubt in many eyes. Numerica had functioned without zero for generations, and the idea of integrating this new concept was daunting for many.

"How can a single symbol make such a difference?" someone shouted from the crowd.

Demonstrating Zero's Power

Lira anticipated this reaction and was prepared. She called upon a few volunteers from different professions to demonstrate the practical applications of zero. Harpreet Kaur, an architect, stepped forward first.

"Imagine trying to construct a building without precise measurements," Harpreet said. "Zero allows us to mark a starting point, ensuring stability and accuracy. With it, our designs can reach new heights."

Next, Baljit Singh, a merchant, took the stage. “In commerce, zero helps us balance our books accurately. Without it, distinguishing between profit and loss is nearly impossible.”

THE SCHOLAR’S CONTRIBUTION

Dr. Amandeep Kaur, a respected scholar, shared her experiences. “Zero simplifies complex equations, making scientific exploration more accessible. It bridges the gap between the positive and the negative, providing a point of balance in our calculations.”

Each demonstration was met with thoughtful nods and murmurs of agreement. Slowly, the skepticism began to melt away, replaced by a growing sense of wonder.

A Turning Point

Lira continued, "Zero is not just a number. It represents the concept of nothingness, a bridge between what is and what isn't. It is a point of balance in the infinite dance of numbers."

The crowd listened intently as Lira explained the philosophical implications of zero. She spoke of how it allowed for the representation of absence and the understanding of infinite possibilities.

"With zero, we can simplify our calculations, bring precision to our measurements, and gain new perspectives on the nature of existence," she concluded.

Overcoming Fear

Despite the compelling arguments, some people still hesitated. The fear of change loomed large. Lira addressed this head-on.

"I understand that change can be frightening," she said gently. "But think about the benefits zero can bring. Imagine a world where our calculations are accurate, our structures are stable, and our knowledge is boundless."

She paused, letting her words sink in. "We have the opportunity to embrace a new era of prosperity and enlightenment. Let us not let fear hold us back."

PRACTICAL APPLICATIONS

Lira then showcased some practical applications of zero that could immediately benefit Numerica. She demonstrated how zero could be used in timekeeping to create more accurate calendars and clocks. Eldrin, the timekeeper, verified her claims, showing the crowd how zero marked the beginning of cycles, making scheduling and time management more efficient.

She also introduced zero-based methods for accounting and bookkeeping, making it easier for merchants to track their transactions and balance their accounts. Baljit Singh shared his experiences, emphasizing how zero had transformed his business practices.

Embracing Change

The demonstrations and explanations began to resonate with the audience. Merchants saw the potential for more accurate record-keeping, architects envisioned more stable and intricate designs, and scholars pondered the new realms of knowledge they could explore.

A prominent merchant, Manjit Singh, stood up. "Lira, your journey has opened our eyes. I see the value of zero in our trade. It's time we embrace this change."

The Role of Tradition

Lira knew that change had to be balanced with respect for tradition. "Numerica's rich history and cultural heritage are not in opposition to zero," she said. "In fact, zero can enhance our traditions, bringing new depth to our understanding and practices."

She proposed integrating zero into existing rituals and ceremonies, showing how it could bring clarity and precision without disrupting the essence of their traditions. This idea was well-received, as people began to see zero not as a threat but as an enhancement.

BUILDING TRUST

Building trust was crucial. Lira encouraged the leaders of Numerica to be transparent about the changes and to involve the community in the transition process. She suggested forming committees to oversee the integration of zero in various sectors, ensuring that everyone had a voice and a role in this new era.

"We must work together, support each other, and embrace this change as a community," Lira said. "Together, we can unlock the full potential of zero."

CELEBRATING PROGRESS

To celebrate the introduction of zero, Lira proposed a grand festival. The festival would not only mark this significant milestone but also educate the public about zero's benefits through interactive exhibits, workshops, and demonstrations.

The idea was met with enthusiasm. Preparations began, and soon the entire city was abuzz with excitement. The festival became a symbol of Numerica's commitment to progress and innovation.

THE FESTIVAL OF ZERO

The Festival of Zero was a resounding success. People from all over Numerica and neighboring regions attended, eager to learn and celebrate. The event featured lectures by scholars, demonstrations by architects and merchants, and interactive sessions for children.

Lira's presentation was the highlight of the festival. She shared her journey, the challenges she faced, and the profound impact of zero on her understanding of the world. Her story inspired many, and by the end of the festival, the acceptance of zero had gained significant momentum.

INTEGRATING ZERO INTO DAILY LIFE

As Numerica embraced zero, its presence became a part of daily life. Schools taught children about zero from an early age, ensuring that future generations would grow up with a deep understanding of its value. Merchants, architects, and scholars integrated zero into their work, reaping the benefits of increased accuracy and efficiency.

Zero-based innovations spread across the city. New timekeeping devices, accounting methods, and architectural designs emerged, each reflecting the transformative power of zero.

THE ROAD TO MASTERY

The journey to fully integrate zero was not without its challenges. There were moments of confusion and resistance, but the community's commitment to learning and adaptation helped them overcome these hurdles.

Lira continued to play a crucial role, offering guidance and support. She mentored young scholars, conducted workshops, and wrote extensively about zero, ensuring that the knowledge was accessible to all.

A NEW IDENTITY

Numerica's identity evolved as it embraced zero. The city became known not only for its mathematical prowess but also for its spirit of curiosity and innovation. Its citizens took pride in their ability to adapt and grow, and Numerica's reputation as a center of learning and progress spread far and wide.

A Catalyst for Collaboration

The acceptance of zero fostered collaboration within Numerica and with neighboring regions. Scholars and leaders from other cities visited Numerica to learn about its integration of zero, leading to the exchange of knowledge and ideas.

Joint projects and research initiatives flourished, resulting in shared advancements that benefited all. Numerica's commitment to progress and cooperation set a new standard, inspiring others to follow suit.

Enhancing Everyday Life

Zero's impact extended to all aspects of life. Farmers used zero-based methods to optimize their planting and harvesting schedules, resulting in better yields. Artisans and craftsmen incorporated zero into their designs, creating more intricate and precise works.

Even in daily interactions, people found ways to apply zero, whether in personal finances, time management, or problem-solving. The concept of zero became a lens through which they viewed the world, unlocking new perspectives and possibilities.

LIRA'S REFLECTIONS

As Numerica thrived, Lira often reflected on her journey. She thought about the challenges she faced, the discoveries she made, and the impact her work had on her city. Her quest for zero had transformed not only her life but the lives of countless others.

Lira's reflections were captured in her writings, which became treasured texts for future generations. Her journals and memoirs provided insight into her thoughts and experiences, inspiring others to embark on their quests for knowledge.

RECOGNIZING CONTRIBUTIONS

Numerica recognized the contributions of those who played pivotal roles in the acceptance of zero. Celebrations and ceremonies honored individuals like Harpreet Kaur, Baljit Singh, Dr. Amandeep Kaur, and many others who had championed the cause.

Lira, in particular, was celebrated as a hero and visionary. Her name became synonymous with curiosity, courage, and the pursuit of knowledge. Statues and monuments were erected in her honor, serving as a reminder of her legacy.

A LASTING LEGACY

Lira's legacy lived on through the institutions she helped establish, the knowledge she shared, and the lives she touched. The Zero Institute continued to be a beacon of learning and innovation, attracting scholars from far and wide who were eager to delve into the mysteries of mathematics and beyond.

INSPIRING FUTURE GENERATIONS

Children grew up hearing tales of Lira's adventures and the profound impact of her discoveries. Schools incorporated her story into their curriculum, ensuring that every child knew the importance of zero and the value of curiosity and perseverance.

One such child, Amrit, was particularly inspired by Lira's journey. He often visited the Zero Institute, soaking up knowledge and dreaming of making his own discoveries. His parents, both scholars, encouraged his passion, and he quickly became a bright star in Numerica's academic community.

Continuing the Quest for Knowledge

Amrit, now a young scholar, embarked on his own journey to explore the mysteries of the universe. He ventured into uncharted territories of mathematics and science, building upon the foundation that Lira had laid. His research led to breakthroughs in understanding infinity, complex numbers, and quantum mechanics.

His work was a testament to Lira's enduring Influence, demonstrating how one person's curiosity could ignite a spark that burned for generations. The discoveries Amrit made were celebrated across Numerica and beyond, further solidifying the city's reputation as a hub of intellectual advancement.

The Expansion of Numerica

As Numerica continued to prosper, the city expanded both in size and influence. New districts were established, each dedicated to different fields of study. The architectural sector flourished under Harpreet Kaur's guidance, incorporating zero-based designs that were both innovative and functional.

Commerce thrived with Baljit Singh's advancements in zero-based accounting, and timekeeping became an art form under Eldrin's meticulous methods. Numerica's economy grew, attracting merchants and traders from distant lands who marveled at the city's precision and organization.

STRENGTHENING COMMUNITY BONDS

The acceptance of zero also strengthened community bonds. People from different backgrounds and professions worked together, united by a shared vision of progress. Public forums and discussion groups became common, where citizens could exchange ideas and collaborate on projects.

These gatherings fostered a sense of belonging and mutual respect, as people realized that their diverse perspectives and skills were crucial to Numerica's success. The city's governance became more inclusive, with representatives from all sectors contributing to decision-making processes.

LIRA'S FINAL DAYS

As Lira grew older, she spent her days reflecting on her journey and mentoring young scholars. She often visited the Zero Institute, where she was greeted with admiration and gratitude. Her presence was a source of inspiration, reminding everyone of the power of curiosity and determination.

One sunny afternoon, Lira passed away peacefully in her sleep. The news of her passing spread quickly, and the entire city mourned the loss of their beloved visionary. A grand memorial service was held at the amphitheater where she had first introduced zero, attended by citizens from all walks of life.

A Memorial to Lira

To honor Lira's memory, Numerica built a grand memorial in the heart of the city. The monument featured intricate carvings and inscriptions detailing her journey and the impact of her discoveries. At the center stood a statue of Lira, holding a scroll with the symbol of zero prominently displayed.

The memorial became a place of pilgrimage for scholars, students, and admirers who came to pay their respects and draw inspiration from Lira's legacy. It was a constant reminder of the transformative power of knowledge and the importance of embracing the unknown.

The Next Chapter

Lira's passing marked the end of an era, but it also signaled the beginning of a new chapter for Numerica. The city continued to thrive, building on the foundations she had laid. Her teachings were carried forward by the next generation of scholars, who were determined to honor her legacy by pushing the boundaries of knowledge.

Amrit, now a leading figure in Numerica's academic community, took on the mantle of guiding the city's intellectual pursuits. He and his peers embarked on ambitious projects, exploring new frontiers and making discoveries that further enriched Numerica's understanding of the universe.

NUMERICA'S GLOBAL INFLUENCE

Numerica's influence extended beyond its borders, inspiring other cities and regions to embrace the concept of zero and adopt similar approaches to learning and innovation. International collaborations flourished, leading to shared advancements that benefited humanity as a whole.

Scholars from around the world visited Numerica to learn from its example, and many chose to stay, contributing to the city's vibrant intellectual community. Numerica became a beacon of hope and progress, demonstrating that with curiosity, courage, and cooperation, anything was possible.

A LEGACY OF CURIOSITY

Lira's story continued to be told for generations, inspiring countless individuals to pursue their own quests for knowledge. Her legacy lived on in the hearts and minds of those who understood the value of asking questions, seeking answers, and never being afraid to explore the unknown.

Her journey from a curious young girl to a revered scholar served as a reminder that the pursuit of knowledge was a lifelong endeavor, one that required dedication, resilience, and an unwavering belief in the power of learning.

THE ETERNAL DANCE OF NUMBERS

As the years passed, Numerica remained a symbol of enlightenment and progress. The city's commitment to knowledge and innovation ensured that it continued to grow and evolve, adapting to new challenges and seizing new opportunities.

Lira's legacy was woven into the very fabric of Numerica, influencing every aspect of life and guiding the city's journey into the future. Her discovery of zero had unlocked a world of infinite possibilities, and Numerica embraced them with open arms.

In the end, Lira's story was not just about zero, but about the enduring power of curiosity, the courage to challenge the status quo, and the profound impact that one person's quest for knowledge could have on the world. And so, the city of Numerica continued its eternal dance of numbers, ever evolving, ever learning, and ever reaching for the star.

CHAPTER

8

VIII. OVERCOMING SKEPTICISM

INITIAL DOUBTS

When Lira first introduced the concept of zero to the people of Numerica, she was met with a wave of skepticism. The citizens had lived their entire lives without zero and were wary of this new, abstract idea. Many believed that their existing system, though flawed, was sufficient and saw no need for such a radical change. There was a deep-rooted resistance to altering traditions that had been in place for generations. The older population, especially, found it difficult to grasp how a single number could have such a profound impact on their way of life. The notion of accepting something they couldn't see or feel seemed ludicrous.

THE TOWN HALL MEETING

To address these concerns, Lira organized a town hall meeting in the grand amphitheater. She invited scholars, merchants, architects, and citizens from all walks of life to come and listen. The amphitheater was packed, and a hum of anticipation filled the air as Lira took the stage. This was a pivotal moment in Numerica's history. The citizens, although skeptical, were curious and eager to hear what Lira had to say. The atmosphere was charged with a mix of doubt and anticipation. Lira knew she had to present her case compellingly to sway public opinion.

LIRA'S PRESENTATION

Lira began by explaining the basic concept of zero and its historical significance. She showed how ancient civilizations had struggled without zero and how its introduction had revolutionized mathematics and science. Using simple examples, she demonstrated how zero could simplify calculations and provide clarity where there was confusion. Her explanations were clear and methodical, aimed at making even the most complex ideas accessible to everyone in the audience. Lira used visual aids and real-life analogies to bridge the gap between abstract theory and practical application, ensuring that her message resonated with all.

REAL-LIFE APPLICATIONS

To make her case more compelling, Lira presented real-life applications of zero. She illustrated how merchants like Baljit Singh could balance their books more accurately, how architects like Harpreet Kaur could design more stable structures, and how timekeepers like Eldrin could achieve precise timekeeping. Each example was met with murmurs of interest and cautious nods from the audience. By showing the tangible benefits of zero, Lira hoped to alleviate some of the fear and uncertainty surrounding its adoption. She stressed that zero was not just a theoretical construct but a practical tool with real-world advantages.

ADDRESSING FEARS

Despite her thorough explanations, there were still those who were resistant to change. Lira addressed their fears directly, acknowledging the challenges and uncertainties that came with adopting a new concept. She reassured them that the transition would be gradual and supported by education and training. Lira understood that change was difficult, especially when it involved something as fundamental as the way people understood numbers. She emphasized that while the transition might be challenging, the long-term benefits would far outweigh the temporary inconveniences.

TESTIMONIALS

To bolster her argument, Lira invited respected members of the community to share their experiences. Rajinder Singh, a scholar who had studied zero extensively, spoke about the mathematical breakthroughs it enabled. Ranjeet Singh, initially skeptical, shared how zero had improved his understanding of commerce and accounting. These testimonials had a profound impact on the audience. Hearing firsthand accounts from trusted figures within their community helped to build trust and credibility. The testimonials served as powerful endorsements of Lira's message, showcasing the real-world benefits of zero through personal stories.

DEMONSTRATIONS

Lira conducted live demonstrations to show zero in action. She balanced complex equations, designed geometric patterns, and simulated timekeeping scenarios, all using zero. The audience watched in awe as the simplicity and power of zero became evident. These demonstrations were crucial in bridging the gap between theory and practice. By seeing zero in action, the citizens of Numerica could visualize its potential impact on their daily lives. The demonstrations helped to demystify zero and showcase its practical applications in a tangible, understandable way.

QUESTIONS AND ANSWERS

The meeting opened up to a Q&A session. Citizens asked numerous questions, ranging from practical applications to philosophical implications. Lira answered each question patiently, providing clear and thoughtful responses that further alleviated doubts. The Q&A session was a vital part of the process, allowing citizens to voice their concerns and get direct answers. It helped to create an open and transparent dialogue, building trust between Lira and the community. By addressing their questions directly, Lira was able to dispel many of the myths and misconceptions surrounding zero.

THE TURNING POINT

The turning point came when Eldrin, the respected timekeeper, stood up and spoke. "I was skeptical at first, but Lira's explanations have opened my eyes. Zero is not just a number; it's a tool that can transform our understanding of the world. I believe it's time we embrace this change." Eldrin's endorsement had a significant impact on the audience. As a respected figure in the community, his support lent credibility to Lira's message and helped to sway public opinion. Eldrin's testimony was a powerful testament to the potential of zero and its transformative impact.

GRADUAL ACCEPTANCE

Eldrin's endorsement had a significant impact. Gradually, more and more people began to see the potential benefits of zero. Small groups formed to discuss and experiment with zero, and Lira provided guidance and resources to support their efforts. The initial wave of acceptance began to spread, with citizens from all walks of life coming together to explore the possibilities that zero offered. This grassroots movement was crucial in building momentum and fostering a sense of collective ownership over the change.

EDUCATIONAL INITIATIVES

To facilitate the transition, Lira and her team launched educational initiatives across Numerica. Workshops, seminars, and classes were organized to teach citizens about zero and its applications. Schools incorporated zero into their curriculum, ensuring that future generations would grow up with a solid understanding of this fundamental concept. Education was a key pillar in the transition process, helping to equip citizens with the knowledge and skills they needed to embrace zero fully. These initiatives played a crucial role in building a foundation for lasting change.

PRACTICAL IMPLEMENTATION

Merchants began using zero in their ledgers, architects started incorporating it into their designs, and timekeepers recalibrated their clocks. The initial challenges were met with determination and ingenuity, and the benefits soon became apparent. The practical implementation of zero marked a significant milestone in Numerica's journey, showcasing its real-world impact. As citizens began to experience the benefits firsthand, their initial skepticism gave way to enthusiasm and support. The success stories that emerged from these practical applications served as powerful endorsements of zero's potential.

COMMUNITY COLLABORATION

The process of integrating zero fostered a sense of community collaboration. Citizens from different professions and backgrounds worked together, sharing their knowledge and experiences. This collaboration led to innovative solutions and strengthened social bonds. The spirit of collaboration and mutual support was a testament to Numerica's resilience and adaptability.

By working together, the community was able to overcome challenges and make significant progress in their journey towards embracing zero. The collaborative efforts also helped to build a sense of unity and shared purpose, strengthening the bonds within the community.

OVERCOMING CHALLENGES

The journey was not without its challenges. There were moments of frustration and setbacks, but Lira's unwavering support and encouragement kept the momentum going. She reminded everyone that progress often required perseverance and the willingness to embrace the unknown. The challenges and setbacks that emerged during the transition process were seen as opportunities for growth and learning. Lira's leadership and guidance were crucial in navigating these challenges and keeping the community motivated and focused on their shared goal.

Success Stories

Success stories began to emerge. Baljit Singh reported a significant improvement in his accounting accuracy, Harpreet Kaur completed a structurally sound and aesthetically pleasing building, and Eldrin's timekeeping methods became the standard across Numerica. These successes reinforced the value of zero and inspired others to follow suit. The success stories that emerged served as powerful testimonials to the transformative impact of zero. They provided tangible evidence of the benefits that zero could bring, inspiring others to embrace the change and contribute to the collective effort.

SHIFTING PERSPECTIVES

As people experienced the practical benefits of zero, their perspectives began to shift. The initial skepticism gave way to curiosity and excitement. Citizens realized that zero was not just a number, but a key to unlocking new possibilities and understanding. The shift in perspectives marked a significant turning point in Numerica's journey, showcasing the transformative power of zero. As citizens began to see the potential and benefits of zero, their enthusiasm and support grew, creating a positive feedback loop that propelled the community forward.

CULTURAL INTEGRATION

Zero became integrated into Numerica's culture. It appeared in art, literature, and even everyday language. Poets and writers found inspiration in the concept of nothingness and infinity, while artists used zero to create stunning visual representations of balance and symmetry. The cultural integration of zero marked a significant milestone in Numerica's journey, showcasing its impact beyond mathematics and science. Zero became a symbol of innovation and progress, inspiring new forms of artistic and intellectual expression.

PHILOSOPHICAL REFLECTIONS

The introduction of zero also sparked philosophical reflections. Scholars and thinkers pondered the deeper meanings of existence, the nature of the universe, and the role of humans in the grand scheme of things. These reflections enriched Numerica's intellectual landscape and opened up new avenues of exploration. The philosophical reflections that emerged from the introduction of zero added depth and complexity to Numerica's intellectual landscape, fostering a culture of curiosity and exploration.

TECHNOLOGICAL ADVANCEMENTS

The acceptance of zero paved the way for technological advancements. Innovations in engineering, astronomy, and computing emerged, propelled by the new mathematical possibilities that zero offered. Numerica became a hub of scientific progress and technological innovation. Scholars and inventors, now free to explore the potential of zero, created devices and systems that revolutionized daily life. Calculations that once took hours could now be completed in minutes, thanks to the newfound efficiency zero brought to mathematics.

EMPOWERING EDUCATION

With zero firmly integrated into Numerica's culture, the education system underwent a transformation. Schools introduced comprehensive curriculums focused on the mathematical and philosophical aspects of zero. Children grew up understanding its significance, learning not only how to use zero in calculations but also appreciating its broader implications. This new generation, empowered with knowledge and curiosity, began to innovate in ways their predecessors could not have imagined. Teachers reported that students showed a remarkable improvement in their problem-solving skills and logical thinking, attributing these advances to the fundamental understanding of zero.

ECONOMIC GROWTH

The economic impact of zero was profound. With better accounting and financial management, businesses flourished. Trade expanded, not just within Numerica but also with neighboring regions. The merchants, who once struggled to balance their books, now experienced unprecedented growth and stability. Economic policies became more effective, guided by accurate data and projections. Numerica's marketplace, which had always been vibrant, now thrived with an unprecedented level of precision and efficiency. Merchants like Baljit Singh expanded their operations, leading to increased employment opportunities and prosperity.

ENGINEERING MARVELS

Architects and engineers, inspired by the precision that zero enabled, began to create structures that were not only aesthetically pleasing but also incredibly stable. Harpreet Kaur, known for her beautiful yet precarious designs, now led projects that were both innovative and safe. The cityscape of Numerica transformed, with towering buildings and intricate bridges that became the envy of neighboring regions. These engineering marvels stood as a testament to the power of zero, showcasing the perfect blend of art and science. Public spaces were redesigned, incorporating zero into their layouts to create harmonious and balanced environments.

ASTOUNDING ASTRONOMICAL DISCOVERIES

Astronomers in Numerica made significant strides, thanks to zero. Eldrin and his team recalibrated their instruments and discovered new celestial bodies and patterns in the night sky. Zero allowed them to chart the stars with unprecedented accuracy, leading to new understandings of the universe. These discoveries sparked a renewed interest in astronomy among the citizens, who marveled at the beauty and complexity of the cosmos. Eldrin's observatory became a center of learning and exploration, attracting scholars and enthusiasts from far and wide. The newfound precision also enabled more accurate predictions of celestial events, enhancing Numerica's reputation as a leader in astronomical research.

ENHANCED TIMEKEEPING

Eldrin's advancements in timekeeping, facilitated by zero, revolutionized the way Numerica tracked time. The recalibrated clocks and calendars brought unparalleled accuracy to daily life. Scheduling events, calculating durations, and planning became much more efficient, leading to better productivity and coordination across the city. The precision in timekeeping also had a positive impact on various industries, from agriculture to transportation, where timing was crucial. Eldrin's innovations set a new standard in timekeeping, with other regions looking to Numerica for guidance and expertise.

PHILOSOPHICAL AND ARTISTIC FLOURISHING

The introduction of zero led to a philosophical renaissance. Thinkers and artists explored the concepts of nothingness and infinity, creating works that reflected these profound ideas. Paintings, sculptures, and literary pieces that incorporated zero captivated audiences, sparking deep reflections on existence and the universe. The cultural scene in Numerica flourished, with zero becoming a central theme in artistic and intellectual discussions. This period of philosophical and artistic flourishing enriched Numerica's cultural heritage, leaving a lasting legacy for future generations.

BRIDGING DIVIDES

The integration of zero also played a role in bridging social divides. The collaborative efforts required to understand and apply zero brought people from different backgrounds together. This spirit of cooperation fostered a sense of unity and shared purpose. Community projects flourished, with citizens working together to improve their city. The sense of unity and collaboration extended beyond Numerica, as the city's success with zero inspired other regions to follow suit. Numerica became a symbol of progress and innovation, demonstrating the power of collective effort and shared vision.

LIRA'S CONTINUED LEADERSHIP

Lira's leadership was instrumental in this transformative journey. Her vision, determination, and ability to inspire others were key to overcoming skepticism and embracing change. She continued to guide Numerica, leading initiatives and supporting innovations that leveraged zero's potential. Her role as a mentor and leader solidified her legacy as a transformative figure in Numerica's history. Lira's continued involvement ensured that the progress made was sustained and built upon, paving the way for future advancements and discoveries.

GLOBAL INFLUENCE

As Numerica flourished, its influence spread beyond its borders. Scholars and leaders from other regions visited to learn about zero and its applications. Numerica's success story became a model for others to follow. International collaborations emerged, leading to a global exchange of ideas and innovations. Numerica's pioneering work with zero set a precedent, inspiring other regions to adopt and integrate zero into their own systems. The city's achievements were celebrated worldwide, cementing Numerica's reputation as a center of knowledge and progress.

LEGACY OF INNOVATION

The legacy of zero in Numerica was not just about the immediate benefits but also about fostering a culture of innovation. The city became a beacon of progress, where new ideas were constantly explored and developed. This spirit of innovation continued to drive Numerica forward, ensuring that it remained at the forefront of scientific and technological advancements. The culture of innovation inspired by zero permeated every aspect of

Numerica's society, encouraging creativity, critical thinking, and a relentless pursuit of knowledge.

A BRIGHT FUTURE

With zero firmly embedded in the fabric of Numerica's society, the city looked towards a bright and promising future. The advancements in various fields laid a solid foundation for continued progress. The citizens, now united by a common understanding and appreciation of zero, were more open to new ideas and innovations. The spirit of curiosity and the willingness to explore the unknown became defining characteristics of Numerica's culture. This forward-thinking mindset positioned Numerica as a leader in the global community, ready to tackle future challenges and seize new opportunities.

CONTINUED EXPLORATION

Encouraged by the success of integrating zero, the scholars of Numerica embarked on new explorations. They delved deeper into mathematical theories, uncovering further nuances and applications of zero. These explorations led to breakthroughs in fields such as physics, computer science, and artificial intelligence. Numerica became a hub for cutting-edge research, attracting the brightest minds from around the world. The city's commitment to continued exploration ensured that it remained at the forefront of scientific and technological advancements, driving progress and innovation.

ECONOMIC PROSPERITY

The economic prosperity that began with the introduction of zero continued to grow. Businesses thrived, trade flourished, and new industries emerged. Numerica's economy diversified, creating a resilient and dynamic economic landscape. The city became a major trade center, with merchants from distant lands eager to engage in commerce. The economic growth also led to improved living standards for the citizens, with better infrastructure, healthcare, and education. Numerica's economic prosperity was a testament to the transformative power of zero, showcasing its impact on both individual and collective well-being.

SOCIAL PROGRESS

With economic stability came social progress. The wealth generated from the flourishing economy was invested in public services and infrastructure. Education, healthcare, and housing saw significant improvements, enhancing the quality of life for all citizens. Social initiatives aimed at reducing inequality and supporting vulnerable populations were implemented, fostering a more inclusive and equitable society. Numerica's commitment to social progress ensured that the benefits of zero were shared by everyone, creating a sense of unity and solidarity within the community.

ENVIRONMENTAL STEWARDSHIP

The people of Numerica also recognized the importance of environmental stewardship. The precision enabled by zero allowed for better management of natural resources and the development of sustainable practices. Innovations in agriculture, energy, and waste management reduced the city's ecological footprint, ensuring a harmonious balance between development and conservation. Environmental stewardship became a core value of Numerica's society, reflecting the city's commitment to preserving the planet for future generations. These efforts not only improved the local environment but also set an example for other regions to follow.

CULTURAL RENAISSANCE

The cultural renaissance sparked by zero continued to flourish. Artists, writers, and musicians drew inspiration from the concept of zero, creating works that explored themes of emptiness, infinity, and balance. Numerica became a vibrant cultural center, known for its rich artistic and intellectual life. Festivals, exhibitions, and performances celebrated the city's heritage and its journey with zero, attracting visitors from around the world.

The cultural renaissance enriched the lives of Numerica's citizens, fostering a sense of pride and identity rooted in the city's innovative spirit.

GLOBAL LEADERSHIP

Numerica's success with zero positioned it as a global leader. The city's scholars and experts were sought after for their knowledge and insights, contributing to international research and development projects. Numerica played a key role in shaping global standards and practices, particularly in the fields of mathematics, science, and technology. The city's leadership extended beyond academia, influencing policies and initiatives aimed at addressing global challenges such as climate change, public health, and economic development. Numerica's global leadership was a testament to the transformative power of zero and the city's commitment to making a positive impact on the world.

INSPIRING FUTURE GENERATIONS

Lira's legacy continued to inspire future generations. Her story was taught in schools, and her achievements were celebrated in annual festivals. Children grew up with role models who exemplified the values of curiosity, perseverance, and the pursuit of knowledge. The spirit of exploration and innovation that Lira embodied became a guiding principle for Numerica's youth, encouraging them to dream big and strive for excellence. The enduring legacy of Lira's journey ensured that Numerica's future remained bright, driven by the passion and determination of its people.

HONORING THE PAST

As Numerica moved forward, it also honored its past. The city's history, including the era before zero, was preserved and celebrated. Museums and historical sites chronicled Numerica's journey, providing valuable lessons and insights for future generations. This respect for the past ensured that the city remained grounded in its heritage while embracing innovation and progress. The preservation of history helped to foster a sense of continuity and identity, reminding the citizens of the challenges they had overcome and the achievements they had made.

A LEGACY OF CURIOSITY

The story of Numerica and zero serves as a lasting legacy of the power of curiosity. It shows that even the smallest and most abstract concepts can lead to profound changes when explored with an open mind and a willingness to embrace the unknown. Lira's journey and Numerica's transformation stand as a testament to the importance of questioning, learning, and innovating. This legacy of curiosity continues to drive Numerica forward, inspiring new generations to push the boundaries of knowledge and understanding.

THE FUTURE AWAITS

As Numerica looked to the future, it did so with confidence and optimism. The city was ready to face new challenges and seize new opportunities, guided by the principles of innovation, collaboration, and curiosity. The journey with zero had shown that with determination and an open mind, anything was possible. Numerica's future was bright, filled with endless possibilities and the promise of continued progress. The city's commitment to exploration and discovery ensured that it would remain a beacon of knowledge and innovation for generations to come. The future awaited, and Numerica was ready to embrace it with open arms.

CHAPTER 9

IX. A NEW ERA

THE DAWN OF TRANSFORMATION

Numerica slowly embraced zero, and with it, a new era of prosperity and enlightenment dawned. The initial skepticism began to fade as the tangible benefits of zero became evident. Citizens started to see how their lives were improving, and the city itself began to transform. Roads were better planned, buildings more stable, and economic transactions smoother. The once unimaginable was now within reach, and Numerica's citizens felt a renewed sense of hope and excitement for the future. This transformation marked the beginning of a new chapter in Numerica's history, one characterized by growth, innovation, and progress.

ECONOMIC STABILITY

With zero integrated into their financial systems, merchants like Baljit Singh and Harpreet Kaur found their accounting much easier to manage. Ledgers balanced accurately, and profit margins were clearer. The marketplace, which had always been vibrant, now operated with unprecedented efficiency. The ability to accurately calculate and track financial data led to more informed business decisions, fostering economic stability and growth. Numerica's economy began to flourish, with new businesses emerging and existing ones expanding. The increased economic stability also led to a rise in employment opportunities, improving the overall quality of life for the citizens.

ARCHITECTURAL BRILLIANCE

Architects, including the renowned Harpreet Kaur, now had the precision they needed to create truly remarkable structures. The city's skyline began to change as new buildings, bridges, and monuments were constructed with an elegance and stability previously thought impossible. These architectural marvels not only enhanced the city's aesthetic appeal but also demonstrated the practical benefits of zero in engineering and design. Public spaces were redesigned to be more functional and beautiful, creating a more pleasant and harmonious environment for the citizens. The advancements in architecture also attracted visitors from neighboring regions, boosting tourism and further stimulating the economy.

ADVANCES IN SCIENCE

Numerica's scholars, now equipped with the concept of zero, made significant strides in various scientific fields. Rajinder Singh and his colleagues at the observatory made groundbreaking astronomical discoveries, charting the stars and planets with unprecedented accuracy. Zero also facilitated advancements in physics and chemistry, leading to new technologies and innovations. The scientific community in Numerica flourished, attracting scholars and researchers from around the world. The increased collaboration and exchange of ideas led to even more discoveries, establishing Numerica as a leading center for scientific research and innovation. The impact of these scientific advancements extended beyond Numerica, influencing global scientific progress.

PRECISION IN TIMEKEEPING

Timekeepers like Eldrin found that zero allowed for unparalleled accuracy in their work. Calendars were more reliable, and clocks kept perfect time. This precision improved everything from daily schedules to agricultural planning. The ability to measure time accurately also had a positive impact on various industries, such as transportation and communication. Eldrin's innovations in timekeeping set a new standard, making Numerica a leader in this field. The improved timekeeping systems were adopted by other regions, enhancing global synchronization and coordination. The advancements in timekeeping also led to new discoveries in astronomy and physics, further advancing Numerica's scientific achievements.

EDUCATIONAL REFORM

The integration of zero into Numerica's educational system brought about a revolution in learning. Schools introduced new curriculums that included the study of zero and its applications.

Children, like Amritjeet, grew up understanding the importance of zero in mathematics and its broader philosophical implications. This new generation of students showed remarkable improvements in their problem-solving skills and critical thinking abilities. The educational reform also encouraged creativity and innovation, nurturing a culture of exploration and curiosity. As a result, Numerica's youth became more engaged and motivated, contributing to the city's ongoing progress and development. The emphasis on education also ensured that Numerica would continue to produce future leaders and innovators.

ENHANCED COMMUNICATION

With zero, Numerica developed more advanced communication systems. The ability to encode and decode information accurately led to the creation of new technologies, such as the telegraph and early computing devices. These innovations revolutionized the way people communicated, making it faster and more efficient. Information could now be transmitted over long distances in a matter of seconds, fostering better connectivity and collaboration. The advancements in communication also facilitated trade and commerce, as businesses could coordinate more effectively. Numerica's leadership in communication technology set a precedent for other regions, further establishing the city's reputation as a hub of innovation.

GLOBAL INFLUENCE

Numerica's success with zero did not go unnoticed. Scholars, traders, and leaders from other regions visited Numerica to learn about their advancements. The city became a model for others to follow, spreading the knowledge and benefits of zero across the globe. This global influence strengthened Numerica's position as a leader in innovation and progress. International collaborations emerged, leading to an exchange of ideas and technologies that benefited all involved. The global recognition also brought economic opportunities, as foreign investments and trade partnerships flourished. Numerica's influence extended beyond its borders, shaping the future of mathematics, science, and technology worldwide.

COMMUNITY ENGAGEMENT

The acceptance of zero fostered a sense of unity and cooperation among Numerica's citizens. Community projects thrived as people worked together to improve their city. The collaborative spirit extended to all aspects of life, from building infrastructure to organizing cultural events. This sense of community engagement strengthened social bonds and created a supportive and inclusive environment. The citizens took pride in their city's achievements and actively participated in its ongoing development. The increased community engagement also led to the emergence of new leaders, who were dedicated to continuing Numerica's legacy of progress and innovation.

ARTISTIC EXPRESSION

The concept of zero inspired a cultural renaissance in Numerica. Artists, writers, and musicians explored themes of nothingness, infinity, and balance in their work. This artistic exploration led to the creation of masterpieces that captivated audiences and sparked deep reflections. Numerica's cultural scene flourished, attracting artists and art enthusiasts from around the world. Festivals, exhibitions, and performances celebrated the city's rich artistic heritage, adding to its vibrant cultural life. The integration of zero into art also fostered a greater appreciation for mathematics and its philosophical implications. This cultural renaissance enriched the lives of Numerica's citizens and contributed to the city's identity as a center of creativity and innovation.

POLITICAL REFORM

The introduction of zero brought about significant changes in Numerica's political landscape. The ability to accurately measure and analyze data led to more informed and effective governance. Policies were based on reliable statistics, leading to better decision-making and resource allocation. The political reform also encouraged transparency and accountability, fostering trust between the government and the citizens. As a result, Numerica's political system became more efficient and responsive to the needs of the people. The improved governance contributed to the city's stability and prosperity, ensuring that the benefits of zero were felt by all.

ENVIRONMENTAL STEWARDSHIP

With zero, Numerica developed better methods for managing natural resources and reducing waste. Innovations in agriculture, energy, and waste management led to more sustainable practices. The city became a model of environmental stewardship, balancing development with conservation. The emphasis on sustainability ensured that Numerica's natural beauty and resources were preserved for future generations. The city's commitment to environmental stewardship also inspired other regions to adopt similar practices, contributing to global efforts to protect the planet. Numerica's achievements in sustainability were celebrated worldwide, reinforcing its reputation as a leader in innovation and progress.

MEDICAL ADVANCEMENTS

The precision enabled by zero led to significant advancements in medicine and healthcare. Doctors and researchers could now make more accurate diagnoses and develop effective treatments. Medical technology improved, leading to better patient outcomes and increased life expectancy. Numerica's healthcare system became one of the best in the world, attracting patients and medical professionals from other regions. The city's commitment to medical advancements also ensured that all citizens had access to high-quality healthcare. The improvements in healthcare contributed to Numerica's overall well-being, allowing the citizens to lead healthier and more fulfilling lives.

TECHNOLOGICAL INNOVATION

Zero opened the door to a new era of technological innovation. Engineers and inventors, inspired by the possibilities of zero, developed groundbreaking technologies that transformed daily life. From computing devices to advanced machinery, these innovations improved efficiency and productivity across various industries. Numerica became a hub of technological development, attracting talent and investment from around the world. The city's leadership in technology set a new standard, driving global progress and innovation. The continuous technological advancements ensured that Numerica remained at the forefront of the digital age, ready to tackle future challenges and opportunities.

SOCIAL PROGRESS

The economic stability and technological advancements brought about by zero led to significant social progress. Numerica's citizens enjoyed improved living standards, better education, and greater access to healthcare. Social initiatives aimed at reducing inequality and supporting vulnerable populations were implemented, fostering a more inclusive and equitable society. The sense of community and cooperation that emerged from the acceptance of zero strengthened social bonds and created a supportive environment. Numerica's commitment to social progress ensured that the benefits of zero were shared by everyone, creating a society where all citizens could thrive.

CULTURAL EXCHANGE

Numerica's success with zero attracted visitors from around the world, leading to a vibrant cultural exchange. Scholars, artists, and traders brought new ideas and perspectives, enriching Numerica's cultural life. This exchange of knowledge and traditions fostered a greater understanding and appreciation of different cultures. Numerica became a melting pot of diversity, where people from various backgrounds lived and worked together harmoniously. The cultural exchange also led to new collaborations and innovations, further enhancing Numerica's reputation as a global center of knowledge and creativity. The city's commitment to cultural exchange ensured that it remained a dynamic and inclusive society.

EDUCATIONAL EXCELLENCE

The integration of zero into Numerica's educational system led to a new era of educational excellence. Schools and universities became centers of innovation, attracting students and scholars from around the world. The emphasis on critical thinking, problem-solving, and creativity nurtured a culture of exploration and curiosity. Numerica's educational institutions produced leaders and innovators who were prepared to tackle the challenges of the future. The city's commitment to education ensured that its citizens were well-equipped with the knowledge and skills needed to drive ongoing progress and development. This emphasis on educational excellence created a virtuous cycle, where each generation built upon the achievements of the previous one, leading to continuous improvement and innovation.

PUBLIC HEALTH INITIATIVES

With a better understanding of zero, Numerica could accurately track and analyze public health data. This led to the development of effective public health initiatives and preventive measures. The city implemented vaccination programs, sanitation improvements, and health education campaigns that significantly reduced the incidence of diseases. The advancements in public health contributed to a higher quality of life and increased life expectancy for Numerica's citizens. The city's proactive approach to public health became a model for other regions, demonstrating the importance of data-driven decision-making in healthcare.

FINANCIAL SYSTEMS

The integration of zero revolutionized Numerica's financial systems. Banks, led by visionary leaders like Gaganpreet Singh, introduced new financial products and services that were previously unimaginable. Interest calculations, risk assessments, and financial modeling became more accurate and sophisticated. The improved financial systems fostered economic growth and stability, attracting investments and creating wealth. Numerica's financial sector became a cornerstone of its economy, providing the necessary support for businesses and individuals to thrive. The advancements in financial systems also promoted financial literacy, empowering citizens to make informed decisions about their finances.

AGRICULTURAL ADVANCEMENTS

The precision enabled by zero led to significant advancements in agriculture. Farmers like Sukhpreet Kaur adopted new techniques and technologies that optimized crop yields and resource utilization. Accurate data collection and analysis allowed for better planning and management of agricultural activities. The improvements in agriculture ensured food security and increased productivity, supporting Numerica's growing population. The city's agricultural advancements also had a positive environmental impact, promoting sustainable farming practices and reducing waste. Numerica's success in agriculture served as a model for other regions, contributing to global efforts to achieve food security and sustainability.

INFRASTRUCTURE DEVELOPMENT

Zero facilitated the development of more efficient and reliable infrastructure. Engineers like Amarjeet Singh designed advanced transportation systems, including railways and roads, that connected different parts of the city and beyond. The improved infrastructure enhanced mobility, reduced travel times, and supported economic activities. The city's investment in infrastructure development also created jobs and stimulated economic growth. Numerica's modern infrastructure became a symbol of its progress and innovation, attracting businesses and residents. The continuous improvements in infrastructure ensured that Numerica remained a dynamic and well-connected city, ready to meet the demands of the future.

CULTURAL RENAISSANCE

The concept of zero inspired a cultural renaissance in Numerica. Artists, writers, and musicians explored themes of nothingness, infinity, and balance in their work. This artistic exploration led to the creation of masterpieces that captivated audiences and sparked deep reflections. Numerica's cultural scene flourished, attracting artists and art enthusiasts from around the world. Festivals, exhibitions, and performances celebrated the city's rich artistic heritage, adding to its vibrant cultural life. The integration of zero into art also fostered a greater appreciation for mathematics and its philosophical implications. This cultural renaissance enriched the lives of Numerica's citizens and contributed to the city's identity as a center of creativity and innovation.

TECHNOLOGICAL INNOVATION

Zero opened the door to a new era of technological innovation. Engineers and inventors, inspired by the possibilities of zero, developed groundbreaking technologies that transformed daily life. From computing devices to advanced machinery, these innovations improved efficiency and productivity across various industries. Numerica became a hub of technological development, attracting talent and investment from around the world. The city's leadership in technology set a new standard, driving global progress and innovation. The continuous technological advancements ensured that Numerica remained at the forefront of the digital age, ready to tackle future challenges and opportunities.

POLITICAL STABILITY

The introduction of zero brought about significant changes in Numerica's political landscape. The ability to accurately measure and analyze data led to more informed and effective governance. Policies were based on reliable statistics, leading to better decision-making and resource allocation. The political reform also encouraged transparency and accountability, fostering trust between the government and the citizens. As a result, Numerica's political system became more efficient and responsive to the needs of the people. The improved governance contributed to the city's stability and prosperity, ensuring that the benefits of zero were felt by all. Numerica's political stability also attracted foreign investments and strengthened diplomatic relations with other regions.

SOCIAL EQUITY

The economic stability and technological advancements brought about by zero led to significant social progress. Numerica's citizens enjoyed improved living standards, better education, and greater access to healthcare. Social initiatives aimed at reducing inequality and supporting vulnerable populations were implemented, fostering a more inclusive and equitable society. The sense of community and cooperation that emerged from the acceptance of zero strengthened social bonds and created a supportive environment. Numerica's commitment to social progress ensured that the benefits of zero were shared by everyone, creating a society where all citizens could thrive.

ENVIRONMENTAL CONSERVATION

Numerica's citizens became more aware of the importance of environmental conservation. With the precision enabled by zero, they developed better methods for managing natural resources and reducing waste. Innovations in agriculture, energy, and waste management led to more sustainable practices. The city became a model of environmental stewardship, balancing development with conservation. The emphasis on sustainability ensured that Numerica's natural beauty and resources were preserved for future generations. The city's commitment to environmental stewardship also inspired other regions to adopt similar practices, contributing to global efforts to protect the planet.

PUBLIC SAFETY

The integration of zero improved Numerica's public safety systems. Accurate data collection and analysis allowed law enforcement agencies to predict and prevent crimes more effectively. Emergency response systems became more efficient, ensuring that help arrived promptly when needed. The advancements in public safety created a secure and peaceful environment for the citizens. Numerica's commitment to public safety also fostered trust between the community and law enforcement agencies, promoting cooperation and mutual respect. The improved public safety systems contributed to the city's overall well-being, allowing the citizens to live without fear and focus on their personal and professional growth.

SCIENTIFIC EXPLORATION

The acceptance of zero enabled Numerica's scholars to make significant strides in scientific exploration. Rajinder Singh and his colleagues at the observatory made groundbreaking astronomical discoveries, charting the stars and planets with unprecedented accuracy. Zero also facilitated advancements in physics and chemistry, leading to new technologies and innovations. The scientific community in Numerica flourished, attracting scholars and researchers from around the world. The increased collaboration and exchange of ideas led to even more discoveries, establishing Numerica as a leading center for scientific research and innovation. The impact of these scientific advancements extended beyond Numerica, influencing global scientific progress.

ECONOMIC GROWTH

With zero integrated into their financial systems, Numerica's economy began to flourish. Merchants found their accounts balanced, architects saw their designs reach new heights, and timekeepers marveled at the accuracy they could achieve. The improved economic stability led to a rise in employment opportunities, improving the overall quality of life for the citizens. The city's economic growth attracted investments and created wealth, further stimulating progress and development. Numerica's economy became a model of efficiency and stability, demonstrating the power of zero in driving economic prosperity. The continuous economic growth ensured that Numerica remained a thriving and dynamic city.

GLOBAL INFLUENCE

Numerica's success with zero did not go unnoticed. Scholars, traders, and leaders from other regions visited Numerica to learn about their advancements. The city became a model for others to follow, spreading the knowledge and benefits of zero across the globe. This global influence strengthened Numerica's position as a leader in innovation and progress. International collaborations emerged, leading to an exchange of ideas and technologies that benefited all involved. The global recognition also brought economic opportunities, as foreign investments and trade partnerships flourished. Numerica's influence extended beyond its borders, shaping the future of mathematics, science, and technology worldwide.

LIRA'S CONTINUED JOURNEY

As Numerica embraced zero and transformed into a beacon of progress and enlightenment, Lira's journey continued. She became a respected figure, her contributions celebrated by the citizens. Lira's quest for knowledge and her determination to share it with others had brought about a profound change in Numerica. However, she knew that her work was far from over. Lira continued to explore new ideas and concepts, driven by an insatiable curiosity and a desire to make the world a better place. Her journey was a testament to the power of curiosity, courage, and the enduring pursuit of knowledge.

THE LEGACY OF ZERO

The story of Numerica and the integration of zero serves as a powerful reminder of the importance of innovation and progress. Lira's journey and the city's transformation demonstrate how a single concept can change the course of history. The legacy of zero and Numerica's achievements continue to inspire future generations, encouraging them to explore new ideas and challenge the status quo. The city's commitment to knowledge, progress, and inclusivity ensures that its legacy will endure, shaping the future and driving continued advancements in mathematics, science, technology, and society.

CHAPTER

10

X. LIRA'S LEGACY

REFLECTIONS ON THE PAST

Lira stood at the edge of the grand amphitheater, her eyes sweeping over the gathered crowd. The amphitheater, once a symbol of Numerica's rich history, now represented its bright future. As she prepared to speak, Lira reflected on her incredible journey. She remembered the initial struggles, the skepticism, and the countless hours of research and exploration. She thought about her grandfather, Thalos, whose wisdom had ignited her curiosity and set her on this path. Lira knew that the journey had not been easy, but the transformation of Numerica was proof of the power of perseverance and knowledge.

THE UNVEILING OF ZERO

With a deep breath, Lira began her speech, recounting the history of zero and its significance. She spoke of the ancient manuscript she had found and the profound impact that zero had on their understanding of mathematics and the universe. As she explained the concept to the crowd, she could see the realization dawning on their faces. Zero was no longer just an abstract idea; it was a symbol of balance, emptiness, and infinite potential. Lira's eloquence and passion captivated the audience, making them see the world through a new lens.

THE PEOPLE'S REACTION

Initially, the people of Numerica had been skeptical of zero. They had lived their entire lives without it, and the idea of embracing such a radical change was daunting. However, as Lira continued her explanation, demonstrating the practical applications and benefits of zero, the skepticism began to fade. Merchants like Baljit Singh and Harpreet Kaur saw how zero could simplify their accounting, while architects like Amarjeet Singh realized its potential for precision in their designs. The crowd's reaction

shifted from doubt to curiosity and finally to acceptance.

PRACTICAL APPLICATIONS

Lira showcased various practical applications of zero, emphasizing its role in simplifying complex calculations and enhancing precision. She demonstrated how zero could revolutionize commerce, engineering, timekeeping, and more. The examples she provided were not just theoretical but rooted in real-world scenarios that the citizens could relate to. Lira's ability to connect abstract concepts with everyday life made the benefits of zero tangible and accessible. The practical applications of zero resonated deeply with the audience, helping them understand its transformative power.

ECONOMIC TRANSFORMATION

The economic impact of zero was profound. With the ability to accurately represent nothing and perform complex calculations, Numerica's merchants and traders experienced a renaissance. Balancing books, understanding profits and losses, and making precise financial decisions became achievable. The newfound accuracy in financial systems fostered economic stability and growth. Merchants like Baljit Singh could now expand their businesses with confidence, knowing that their financial records were accurate and reliable. The economic transformation brought about by zero created a ripple effect, boosting Numerica's prosperity and attracting investments from far and wide.

ARCHITECTURAL INNOVATIONS

Architects like Amarjeet Singh found that zero brought a new level of precision to their work. Measurements and proportions that were once guesswork became exact. This precision led to the construction of more stable and aesthetically pleasing structures. The city's skyline began to change, with buildings reaching new heights of architectural excellence. The integration of zero into architectural practices also sparked a wave of creativity, as architects explored innovative designs that were previously impossible. Numerica's architectural innovations became a symbol of its progress and ingenuity.

ADVANCEMENTS IN TIMEKEEPING

Timekeeper Eldrin marveled at the accuracy that zero brought to timekeeping. The ability to mark the beginning of cycles with zero allowed for precise scheduling and time calculations. Numerica's calendars and clocks became more reliable, and the city's daily life flowed more smoothly. Events were scheduled with confidence, and the synchronization of activities improved efficiency. The advancements in timekeeping also had a positive impact on scientific research, as precise time measurements were crucial for experiments and observations. Numerica's commitment to accurate timekeeping became a cornerstone of its progress.

SCIENTIFIC DISCOVERIES

Zero's introduction led to groundbreaking scientific discoveries. Scholars like Rajinder Singh and his colleagues delved into the mysteries of the universe with newfound precision. Astronomical observations became more accurate, leading to discoveries that expanded humanity's understanding of the cosmos. In fields like physics and chemistry, zero enabled more precise measurements and calculations, resulting in new technologies and innovations. Numerica's scientific community thrived, attracting researchers from around the world and fostering a culture of curiosity and exploration. The scientific discoveries made possible by zero had a lasting impact, shaping the future of science and technology.

EDUCATIONAL REFORMS

Numerica's educational system underwent significant reforms to incorporate the concept of zero. Schools and universities embraced the new knowledge, ensuring that students were well-versed in its applications. Educational institutions focused on developing critical thinking and problem-solving skills, preparing students for a rapidly changing world. The emphasis on education created a generation of innovators and leaders who were equipped to drive Numerica's progress. The city's commitment to education also fostered a culture of lifelong learning, encouraging citizens to continually seek knowledge and self-improvement. The educational reforms laid the foundation for sustained growth and innovation.

CULTURAL RENAISSANCE

The concept of zero inspired a cultural renaissance in Numerica. Artists, writers, and musicians explored themes of nothingness, infinity, and balance in their work. This artistic exploration led to the creation of masterpieces that captivated audiences and sparked deep reflections. Numerica's cultural scene flourished, attracting artists and art enthusiasts from around the world. Festivals, exhibitions, and performances celebrated the city's rich artistic heritage, adding to its vibrant cultural life. The integration of zero into art also fostered a greater appreciation for mathematics and its philosophical implications. This cultural renaissance enriched the lives of Numerica's citizens and contributed to the city's identity as a center of creativity and innovation.

SOCIAL IMPACT

The social impact of zero was significant. Numerica's citizens enjoyed improved living standards, better education, and greater access to healthcare. Social initiatives aimed at reducing inequality and supporting vulnerable populations were implemented, fostering a more inclusive and equitable society. The sense of community and cooperation that emerged from the acceptance of zero strengthened social bonds and created a supportive environment. Numerica's commitment to social progress ensured that the benefits of zero were shared by everyone, creating a society where all citizens could thrive.

PUBLIC HEALTH ADVANCEMENTS

With a better understanding of zero, Numerica could accurately track and analyze public health data. This led to the development of effective public health initiatives and preventive measures. The city implemented vaccination programs, sanitation improvements, and health education campaigns that significantly reduced the incidence of diseases. The advancements in public health contributed to a higher quality of life and increased life expectancy for Numerica's citizens. The city's proactive approach to public health became a model for other regions, demonstrating the importance of data-driven decision-making in healthcare.

INFRASTRUCTURE IMPROVEMENTS

Zero facilitated the development of more efficient and reliable infrastructure. Engineers like Amarjeet Singh designed advanced transportation systems, including railways and roads, that connected different parts of the city and beyond. The improved infrastructure enhanced mobility, reduced travel times, and supported economic activities. The city's investment in infrastructure development also created jobs and stimulated economic growth. Numerica's modern infrastructure became a symbol of its progress and innovation, attracting businesses and residents. The continuous improvements in infrastructure ensured that Numerica remained a dynamic and well-connected city, ready to meet the demands of the future.

POLITICAL REFORMS

The introduction of zero brought about significant changes in Numérica's political landscape. The ability to accurately measure and analyze data led to more informed and effective governance. Policies were based on reliable statistics, leading to better decision-making and resource allocation. The political reform also encouraged transparency and accountability, fostering trust between the government and the citizens. As a result, Numerica's political system became more efficient and responsive to the needs of the people. The improved governance contributed to the city's stability and prosperity, ensuring that the benefits of zero were felt by all. Numerica's political stability also attracted foreign investments and strengthened diplomatic relations with other regions.

TECHNOLOGICAL INNOVATION

Zero opened the door to a new era of technological innovation. Engineers and inventors, inspired by the possibilities of zero, developed groundbreaking technologies that transformed daily life. From computing devices to advanced machinery, these innovations improved efficiency and productivity across various industries. Numerica became a hub of technological development, attracting talent and investment from around the world. The city's leadership in technology set a new standard, driving global progress and innovation. The continuous technological advancements ensured that Numerica remained at the forefront of the digital age, ready to tackle future challenges and opportunities.

ENVIRONMENTAL CONSERVATION

Numerica's citizens became more aware of the importance of environmental conservation. With the precision enabled by zero, they developed better methods for managing natural resources and reducing waste. Innovations in agriculture, energy, and waste management led to more sustainable practices. The city became a model of environmental stewardship, balancing development with conservation. The emphasis on sustainability ensured that Numerica's natural beauty and resources were preserved for future generations. The city's commitment to environmental stewardship also inspired other regions to adopt similar practices, contributing to global efforts to protect the planet.

PUBLIC SAFETY IMPROVEMENTS

The integration of zero improved Numerica's public safety systems. Accurate data collection and analysis allowed law enforcement agencies to predict and prevent crimes more effectively. Emergency response systems became more efficient, ensuring that help arrived promptly when needed. The advancements in public safety created a secure and peaceful environment for the citizens. Numerica's commitment to public safety also fostered trust between the community and law enforcement agencies, promoting cooperation and mutual respect. The improved public safety systems contributed to the city's overall well-being, allowing the citizens to live without fear and focus on their personal and professional growth.

GLOBAL INFLUENCE

Numerica's success with zero did not go unnoticed. Scholars, traders, and leaders from other regions visited Numerica to learn about their advancements. The city became a beacon of knowledge and innovation, its influence spreading far and wide. Diplomatic relations were strengthened as Numerica shared its discoveries and technologies with the world. The exchange of ideas and resources fostered global cooperation and progress. Numerica's influence extended beyond mathematics and commerce; it became a symbol of what could be achieved through curiosity, perseverance, and a willingness to embrace new ideas.

CULTURAL EXCHANGE

As Numerica opened its doors to the world, a rich cultural exchange flourished. People from different backgrounds brought their traditions, languages, and arts, enriching Numerica's cultural tapestry. Festivals celebrating diverse cultures became a regular occurrence, fostering mutual understanding and respect. The exchange of cultural ideas sparked creativity and innovation in various fields, from the arts to science. This cultural exchange also led to the development of new culinary traditions, fashion trends, and artistic styles, further enhancing Numerica's vibrant and dynamic society.

EDUCATIONAL OUTREACH

Numerica's scholars, inspired by Lira's journey, embarked on missions to spread the knowledge of zero to other regions. They established schools and universities, ensuring that the next generation of thinkers and innovators had access to the revolutionary concept. Educational outreach programs provided resources and training to communities that had been isolated from such knowledge. These initiatives helped bridge educational gaps and promote equal opportunities for learning. The widespread dissemination of zero's knowledge led to a global renaissance in mathematics and science, empowering individuals to contribute to their communities' development and progress.

ECONOMIC PROSPERITY

With the integration of zero, Numerica's economy flourished. Accurate financial systems and innovative business practices attracted investors and entrepreneurs. The city became a hub of commerce and trade, with markets bustling with activity. Numerica's economic prosperity improved the standard of living for its citizens, creating jobs and opportunities for growth. The wealth generated was invested in public infrastructure, healthcare, and education, further enhancing the city's development. Numerica's economic model became a blueprint for other regions, demonstrating the power of integrating zero into financial systems.

HEALTHCARE ADVANCES

The precision enabled by zero revolutionized Numerica's healthcare system. Accurate data collection and analysis improved diagnostic tools and treatment methods. Medical research flourished, leading to breakthroughs in various fields, including genetics, pharmacology, and surgical techniques. The advancements in healthcare extended life expectancy and improved the quality of life for Numerica's citizens. Public health campaigns and preventive care programs became more effective, reducing the incidence of diseases. Numerica's healthcare system became a model of excellence, attracting medical professionals and researchers from around the world.

TECHNOLOGICAL BREAKTHROUGHS

Zero's introduction sparked a wave of technological breakthroughs. Inventors and engineers developed cutting-edge technologies that transformed everyday life. From advanced computing systems to automation and robotics, these innovations enhanced efficiency and productivity across all sectors. Numerica became a leader in technological development, driving progress in fields such as artificial intelligence, renewable energy, and space exploration. The continuous stream of technological advancements ensured that Numerica remained at the forefront of global innovation, ready to tackle future challenges and opportunities.

SCIENTIFIC RESEARCH

The acceptance of zero opened new frontiers in scientific research. Numerica's scientists made significant discoveries in physics, astronomy, and biology. The precision afforded by zero allowed for more accurate experiments and observations, leading to groundbreaking insights. Numerica became a center of scientific excellence, attracting researchers from around the world. Collaborative projects and international conferences fostered the exchange of ideas and knowledge, accelerating scientific progress. The discoveries made in Numerica contributed to a deeper understanding of the universe and the development of new technologies that benefited humanity as a whole.

INFRASTRUCTURE DEVELOPMENT

The integration of zero facilitated the development of sophisticated infrastructure projects. Engineers and urban planners designed efficient transportation systems, sustainable housing, and advanced public utilities. The city's infrastructure supported economic growth and improved the quality of life for its residents. Innovations in construction techniques and materials led to the creation of durable and environmentally friendly buildings. The continuous improvements in infrastructure ensured that Numerica could accommodate its growing population and remain a thriving and well-connected city. The city's commitment to infrastructure development also attracted international investments and collaborations, further boosting its economic and technological capabilities.

ENVIRONMENTAL STEWARDSHIP

Numerica's embrace of zero also led to a heightened awareness of environmental stewardship. The precision in data collection and analysis allowed for better resource management and conservation efforts. Sustainable practices were implemented in agriculture, industry, and urban development. The city invested in renewable energy sources, reducing its carbon footprint and promoting a healthier environment. Community initiatives focused on recycling, waste reduction, and the preservation of natural habitats. Numerica's commitment to environmental stewardship set an example for other regions, demonstrating the importance of balancing development with the protection of natural resources.

PUBLIC SAFETY ENHANCEMENTS

The accurate data enabled by zero improved Numerica's public safety systems. Law enforcement agencies could predict and prevent crimes more effectively, ensuring a safer environment for citizens. Emergency response services became more efficient, with faster and more coordinated actions during crises. The advancements in public safety fostered a sense of security and trust within the community. Citizens felt confident in their daily lives, knowing that the city was committed to their well-being. Numerica's public safety enhancements became a model for other regions, showcasing the benefits of integrating zero into security and emergency management systems.

LEGAL AND JUDICIAL REFORMS

The introduction of zero also brought about significant legal and judicial reforms. Accurate record-keeping and data analysis improved the administration of justice. Legal processes became more transparent and efficient, ensuring fair and timely resolutions of cases. The reforms strengthened the rule of law, fostering trust in the judicial system. Numerica's commitment to justice and equality created a more inclusive and harmonious society. The legal and judicial reforms also attracted legal scholars and practitioners, enhancing the city's reputation as a center of legal excellence.

INTERNATIONAL COLLABORATIONS

Numerica's success with zero led to numerous international collaborations. The city partnered with other regions on research projects, technological development, and cultural exchanges. These collaborations fostered mutual understanding and shared progress. Numerica became a key player on the global stage, contributing to international efforts in areas such as climate change, public health, and scientific research. The city's international collaborations also opened new markets for its businesses and provided opportunities for cultural and intellectual exchange. Numerica's role in fostering global partnerships highlighted the importance of cooperation and the benefits of sharing knowledge and resources.

LIRA'S LEGACY

As Numerica thrived, Lira's legacy became an integral part of its identity. Her journey from a curious young girl to a visionary leader inspired countless individuals. Schools, libraries, and institutions were named in her honor, preserving her contributions for future generations. Lira's story was told and retold, a reminder of the power of curiosity, perseverance, and the courage to explore the unknown. Her legacy lived on, teaching generations that sometimes, the smallest and most unassuming symbols can hold the greatest significance. Lira's impact extended beyond Numerica, influencing thinkers, innovators, and leaders around the world. Her dedication to knowledge and progress became a guiding light, inspiring others to pursue their dreams and make a difference.

A BRIGHT FUTURE

With the acceptance of zero, Numerica entered a new era of prosperity and enlightenment. The city's commitment to knowledge, innovation, and cooperation ensured a bright future. Numerica's citizens, united by a shared vision, continued to build on the foundations laid by Lira and her predecessors. The city's progress and achievements became a testament to the transformative power of zero and the enduring spirit of exploration. Numerica's bright future was a reflection of its past, a journey of discovery and growth that continued to inspire and lead the world into new horizons. The legacy of Lira and the significance of zero remained at the heart of Numerica's identity, guiding its path forward and ensuring that the city remained a beacon of knowledge and progress for generations to come.

THE END

And so, the story of Numerica and Lira serves as a reminder of the power of curiosity and the courage to explore the unknown. Lira's legacy lived on, teaching generations that sometimes, the smallest and most unassuming symbols can hold the greatest significance.

SUMMARY

Chapter 1: The Missing Link

In the ancient city of Numerica, every aspect of life was governed by mathematics. However, without zero, the citizens faced numerous challenges in commerce, science, timekeeping, and engineering.

Chapter 2: Commerce in Chaos

Merchants like Jaspreet Kaur struggled to keep accurate accounts without zero, making financial management chaotic and confusing.

Chapter 3: The Architect's Dilemma

Architect Harpreet Singh found that without zero, his designs lacked precision and stability, affecting the quality and safety of his buildings.

Chapter 4: The Scholar's Tale

Lira Kaur's grandfather, Thalos, explained how zero was introduced long ago by a wandering sage but was banished due to fear of its power, leading to its loss over time.

Chapter 5: The Timekeeper's Struggle

Timekeeper Rajveer Kaur faced immense challenges in tracking time accurately without zero, affecting the city's schedules and calendars.

Chapter 6: Lira's Quest

Determined to uncover the secrets of zero, Lira embarked on a journey, meeting scholars, merchants, architects, and timekeepers who shared their struggles.

Chapter 7: The Revelation

Lira returned to Numerica and shared her discoveries, demonstrating how zero could simplify complex calculations, bring precision to measurements, and offer new perspectives.

Chapter 8: Overcoming Skepticism

Initially skeptical, the citizens began to see the potential of zero as Lira provided hands-on examples, transforming the lives of merchants, architects, and timekeepers.

Chapter 9: A New Era

Numerica embraced zero, leading to economic prosperity, architectural excellence, and timekeeping perfection, united in the pursuit of knowledge and progress.

Chapter 10: Lira's Legacy

Lira's legacy lived on, teaching generations the power of curiosity and the courage to explore the unknown. Her contributions transformed Numerica into a hub of innovation and learning.

The End

i

2024 Edition

i A Sardargarh Inc Books Publications

www.ingramcontent.com/pod-product-compliance
Lightning Source LLC
LaVergne TN
LVHW102009210926
848483LV00010B/206